AF531629

Breeding of Fruits and Plantation Crops

NIPA GENX ELECTRONIC RESOURCES & SOLUTIONS P. LTD.
New Delhi-110 034

About the Authors

Prof. Bhimasen Naik (b.1961) hails from an agri-horticultural family of Western Odisha (Village: Jamtalia, Tehsil: Sundargarh, District: Sundargarh). He obtained B.Sc. (Ag. & A.H.) and M.Sc. (Ag.) in Genetics and Plant Breeding from Chandra Shekhar Azad University of Agriculture and Technology, Kanpur in 1981 and 1983, respectively, and Ph.D. from Utkal University, Bhubaneswar in 1999. He was been awarded ICAR Junior Research Fellowship and CSIR Senior Research Fellowship during his studies. He is a Fellow of the Indian Society of Genetics & Plant Breeding, New Delhi; and life member of the Crop Improvement Society of India, Ludhiana; the Indian Society of Plant Breeders, Coimbatore; the Society for Advancement of Rice Research, Hyderabad; the Indian Society of Oilseeds Research, Hyderabad and the Society for Plant Biochemistry & Biotechnology, New Delhi. He has worked on breeding and genetics of ginger, turmeric, mungbean, linseed and rice. He has more than 36 years of research and teaching experience at Odisha University of Agriculture and Technology. He has 56 research papers published in peer-reviewed national and international journals, 23 abstracts of symposia, 7 books, 1 book chapter (Springer-Nature) and 20 popular articles to his credit. He has guided one M.Sc. (Ag.) and one Ph.D. student. He is associated in the development of five cultivars in ginger (Suruchi), turmeric (Surama, Ranga and Rashmi) and linseed (Arpita); collaborated in development of three cultivars in rice (Ashutosh, Gobinda and Hasanta) and registered one genetic stock of mungbean (BSN 1, INGR 00011).

Dr. Ranjan Kumar Tarai (b.1977) obtained his B.Sc. (Ag.) in 1999 from College of Agriculture, Chiplima under Odisha University of Agriculture and Technology; M.Sc. (Hort.) in Fruits and Orchard Management and Ph.D. (Hort.) in Fruits and Orchard Management in 2002 and 2006 respectively from Bidhan Chandra Krishi Viswa Vidyalaya, Mohanpur, Nadia, West Bengal. He successfully qualified NET in Fruit Science during 2004 conducted by ASRB, New Delhi. He started his career in 2006 as Subject Matter Specialist (Horticulture) at Krishi Vigyan Kendra (Gajapati) and later on joined as Junior Scientist (Horticulture) at Regional Research and Technology Transfer Station, Bhawanipatna and as Programme Coordinator at KVK, Kalahandi under Odisha University of Agriculture and Technology, Bhubaneswar. Dr.Tarai is basically a Pomologist inclined to fruit research and teaching. He has an experience of over 14 years of teaching, research and extension. He has published more than 54 research papers in national and international journals, 8 books, 26 popular articles, 18 book chapters, 20 extension booklets & Newsletters to his credit. He has guided four M.Sc. Ag. (Hort.) students. He is a life member of several societies in the field of Horticulture. He has devoted his works in development and transfer of technologies in fruits (especially minor fruit crops), vegetables and horticulture-based farming system suitable for resource poor area. Currently he is working as Associate Professor (Fruit Science) at College of Horticulture, Chiplima under Odisha University of Agriculture and Technology, Bhubaneswar.

Breeding of Fruits and Plantation Crops

Bhimasen Naik
Former Professor of Plant Breeding and Genetics
Odisha University of Agriculture and Technology
Chiplima Campus, Sambalpur-768 025, Odisha

Ranjan Kumar Tarai
Associate Professor of Fruit Science
Odisha University of Agriculture and Technology
Chiplima Campus, Sambalpur-768 025, Odisha

NIPA GENX ELECTRONIC RESOURCES & SOLUTIONS P. LTD.
New Delhi-110 034

NIPA GENX ELECTRONIC RESOURCES & SOLUTIONS P. LTD.

101,103, Vikas Surya Plaza, CU Block
L.S.C.Market, Pitam Pura, New Delhi-110 034
Ph : +91 11 27341616, 27341717, 27341718
E-mail:newindiapublishingagency@gmail.com
www: www.nipabooks.com

For customer assistance, please contact
Phone: + 91-11-27 34 17 17 Fax: + 91-11- 27 34 16 16
E-Mail: feedbacks@nipabooks.com

ISBN: 978-93-94490-33-8

Composed and Designed by NIPA.

To
Late Balamukund Naik
(Grandfather)
and
Sri Purna Chandra Tarai
(Father)

Bhimasen
Ranjan

Preface

The Fifth Deans' Committee of Indian Council of Agricultural Research has recently revised the syllabus of B.Sc. (Hons.) Horticulture which is uniform throughout the country. The course 'Breeding of Fruit and Plantation Crops' is taught in the fourth semester. The present textbook covers the entire syllabus in 24 chapters. Simple and lucid language has been used for easy understanding of the beginners. Attempts have been made to provide latest information; still some valuable information might have been missed. Questions are set at the end of each chapter to assess the understanding of the students.

We have tried our best to remove the errors, typographical or otherwise, from the text; still there might be some. We would highly appreciate if they are brought to our notice for rectification in the next edition. We cherish the encouragement and cooperation received from our family members during preparation of the manuscript. We congratulate NIPA, New Delhi for their support and publishing in a short time.

Though the book is primarily written for B.Sc. (Hons.) Horticulture students, the counterparts of B.Sc. (Hons.) Agriculture also may be benefitted. It may serve as a help book for post-graduate students.

We seek suggestions from learned teachers and students for further improvement of the book.

Authors

Contents

SECTION I

BASIC CONSIDERATIONS

1

An Overview

1.1 Introduction

Fruit breeding is an ancient technology with dynamic current technology and an exciting future (Janick and Moore 1975, 1996). In its broadest sense, fruit breeding refers to the purposeful genetic improvement of fruit crops through various techniques including selection, hybridization, mutation induction and molecular techniques. Its origins trace to the domestication process in prehistory and antiquity, where useful species were chosen and cultivated, and improved by continuous selection (Janick 2005, 2011). Through the millennia genetic improvement of these species have been achieved by grower selection, first from natural seedling populations and then from grower field that continued unique genotypes fixed by vegetative propagation. Spontaneous hybridization between wild plants and cultivated clones was critical to the early domestication of fruits.

1.2 History of Fruit Breeding

"Origin of fruit culture can be traced back to 8000-5000 years". The horticultural practices such as irrigation, vegetative propagation, selection etc. developed about 5000-3000 years ago. The sophisticated fruit culture might have originated 3000-2000 years ago. *Basically, all the cultivated fruit crops are the gift of diversity and not the results of systematic breeding.* Most of them have spread to different parts of the world through various media from the primary centres of origin.

1.2.1 Centres of Origin

The centres of origin of major fruit crops are:

Mediterranean basin : date palm, fig, grape, pomegranate

East Asia : banana, citrus, mango, peach, kiwi fruit, persimmon

Central Asia : pome fruits, apricot

Europe : plum, raspberry

North and South America: avocado, black berry, papaya, pineapple, strawberry

1.2.2 Domestication

In the late Neolithic and Bronze Ages (6000 and 3000 BCE), the ancient Mediterranean fruits (date, olive, grape, fig and pomegranate) were domesticated. Citrus, banana, various pome fruits (apple, pear, quince) and stone fruits (almond, apricot, cherry, peach and plum) were domesticated in Central and East Asia and reached the West in antiquity. A number of fruits and nuts were domesticated only in the 19th and 20th centuries (blue berry, black berry, pecan nut and kiwi fruit). Some genetic changes occurred in a number of fruit crops during domestication. For examples,

a) Breakdown of dioecy in fig, grape, papaya and strawberry

b) Loss of self-incompatibility in cherry

c) Parthenocarpy and seedlessness in apple, pear, banana and plantains, citrus, fig, grapes, loquat, pine apple, persimmon

d) Allopolyploidy in banana and plantain, black berry, raspberry, strawberry, European plum

e) Triploidy in banana and plantain, apple, pear

f) Tetraploidy in raspberry, kiwi fruit, blackberry, blue berry

g) Hexaploidy in kiwifruit, European plum

h) Octaploidy in strawberry,

i) Ease of vegetative propagation in apple rootstocks, date palm

j) Loss of spines, thorns and pubescence in pine apple, citrus, apple, peach, pear

1.2.3 Fruit Breeding

Fruit breeding was elevated as an organized activity during 19th century with reference to mass selection efforts in strawberry and pear. Thomas Andrew Knight (1759-1838) was the first to improve fruits by selection from genetic recombination derived from inter-pollinations of clones. He is an early proponent of the development of plant improvement through cross breeding and selection. He literally initiated the field of fruit breeding. He released a number of improved fruit cultivars (apple, pear, red currant, strawberry, cherry, nectarine and plums). In the USA, fruit breeding became a part of research at the state and federal experiment stations. Many breeding programmes were initiated throughout the USA. Fruit breeding also became an activity of the private sectors. An American nursery man, Luther Burbank (1849-1926) was the first to consider fruit breeding as a commercial endeavour.

Although fruit breeding has been a major activity since the early 20th century, the results have been variable and vary from ineffectual to extraordinarily successful. Progress from breeding a number of fruit crops, however, has shown significant advances in the second half of the 20th century and selections from controlled crosses became significant in many crops. Many of the fruit cultivars are ancient and based on grower-selected seedlings and somatic mutations. Advances in molecular genetics may be the solution for limitations of conventional fruit breeding based on sexual recombination by increasing the selection efficiency using molecular markers and by transgenic technology. Introduction has played a major role in fruit improvement.

1.2.3.1 Indian Context

India is an important centre of origin and diversity of fruit crops like mango, citrus and banana. In 1890, Abdul Baquer Khan introduced grape cultivar 'Anad-e-Shahi' from the Middle East and it revolutionized grape cultivation in the country. It can be considered as the earliest work of the fruit crop improvement. Fruit research in India started in the Department of Botany of six Agricultural Colleges established at Pune, Coimbatore, Lyallpur, Nagpur, Sabour and Kanpur. Establishment of a Division of Horticulture at IARI in 1956-57 and creation of a separate division of Fruits and Horticultural Technology during 1982 boosted the research on different aspects of fruit crops. In 1968, Indian Institute of Horticultural Research (IIHR) at Bangalore and the Central Mango Research Station at Lucknow in 1972 were established and systematic research on fruit crops such as mango, grapes, banana, papaya, guava, sapota, citrus, pineapple etc. was initiated. Plant introduction facilitates further breeding programmes in crops. In India, National Bureau of Plant Genetic Resources (NBPGR) plays the role of the nodal centre for the introduction and maintenance of germplasm of agricultural and horticultural crops including fruits. The State Agricultural Universities and the All India Coordinated Fruit Improvement Project, different National Research Centres under ICAR and other national institutes and sub-centres were responsible for breeding work on all the important fruit crops and have come out with considerable achievements in this field as evident by the number of specific selections and hybrids evolved for commercial cultivation.

1.3 Importance of Fruit Breeding in Fruit Production

Fruit production is the function of cultivar, soil, crop management and crop protection management. Cultivar is the most important input in fruit production. Whatever crop management and plant protection measures are taken up, the fruit production will not be high and of good quality unless the cultivar has the potential. The cultivar is the product of fruit breeding. Through fruit breeding, cultivars may be developed which are yielding high with good quality. Stress resistance may be incorporated into the cultivar. As a result, the expenditure towards management of insect pests and diseases will be minimized. It will decrease the cost of cultivation, thereby increasing the net profit.

1.4 Variability for Economic Traits

In fruit crops, a lot of variability is observed on various economic traits. So, the selection of genotypes with the trait of interest is very crucial step in fruit breeding. Let us see the important economic traits of some of the common fruits.

Mango: Biotic stress resistance for hopper, tip borers, seed borers, powdery mildew, anthracnose, bacterial canker; abiotic stress resistance for cold, water logging, calcium, calcareous soil etc.; dwarfness, high sex ratio, self-incompatibility, red peel colour, high TSS, good keeping quality, regular bearing, early maturity, pickle processing, free from stones etc.

Banana: Resistance to burrowing nematode, Panama wilt, bunchy top, *Fusarium* wilt, leaf spot etc.; good keeping quality, long distance transport, high sugar content, good pulp quality.

Citrus: Resistance to Tristeza virus, gummosis, greening etc., rootstocks for good yield, quality etc.

Guava: Pink skin colour, pink flesh colour, pink skin and flesh, juice making, soft seededness, high TSS, dwarf rootstock, good keeping quality etc.

Grapes: Resistance to *Phylloxera vertifolia*, aphids, thrips, nematodes, powdery mildew, downy mildew, anthracnose, crown gall, *Cercospora* leaf spot, etc. cold, frost, drought, soil salinity; high yield, earliness, basal fruit bud, high TSS, seedlessness, juice, wine, resins etc.

Papaya: Resistance to root rot, viral disease, nematode, cold, frost, water logging, strong wind; good keeping quality, gynodioecious, dwarfness, high TSS, bigger size fruits, red flesh, high papain content etc.

Pineapple: Resistance to mealy ug, wilt, gummosis, root rot, big fruit size, good flavour, early flowering, early ripening etc.

Pomegranate: Tolerance to fruit cracking, good fruit quality and yield, red skin, soft seed etc.

Apple: Resistance to scab, powdery mildew, wooly apple aphid, canker, collar rot, low chilling, drought; high yield, early season, mid-season, late season, red colour, yellow colour, green colour, good shelf life, dwarfness etc.

1.5 Major Problems of Fruit Breeding

- It has long generation cycle of 2-10 years depending upon species and cultivars.
- It is not possible to have more recombination.
- Fruit crop has long juvenile period for which early assessment of strains is not possible, e.g., mango, jackfruit etc.
- Majority of the fruit species are highly heterozygous. So, large population is required for selection.
- Most of the fruit species are polyploid in nature, e.g., banana.
- Polyembryony, e.g., mango, citrus.
- Parthenocarpy and seedlessness, e.g., banana, pineapple.
- Presence of sexual incompatibility, e.g., mango, apple, pear, loquat etc.
- More number of chromosomes creating a big problem in genetical analysis, e.g., ber, mulberry.
- Excessive fruit drop, e.g., mango, citrus, grape etc.
- Presence of single seed in single fruit, thus requiring more number of crosses, e.g., mango, litchi, mahua etc.

1.6 Objectives of Fruit Breeding

The objectives of fruit breeding vary depending upon fruit crops in relation to different locations and requirements. Still the main objective of fruit breeding is to get maximum quality production per unit area with low cost and free from biotic and abiotic stresses. The objectives may also vary with respect to breeding for rootstock and scion.

1.6.1 Objectives for Rootstock

- Wide geographical adaptability.
- Easily propagated, preferably through sexual means.
- It should be compatible with most of the scion cultivars having strong scion-stock union and more longevity.
- It should be resistant to biotic and abiotic stresses.
- For high density planting, it should have dwarfing effect without affecting the productivity of scion cultivars.

- Plant should not have brittle root system.
- It should be free from suckering habit.

1.6.2. Objectives for Scion Cultivars

- Dwarf growth stature.
- Regular, precocious and prolific bearer per unit canopy area.
- High productivity with good quality fruits.
- Resistant to biotic and abiotic stresses.
- Attractive fruit colour with pleasant aroma.
- Suitable for processing and export.
- Good keeping and transport quality.

1.7 Steps in Fruit Breeding

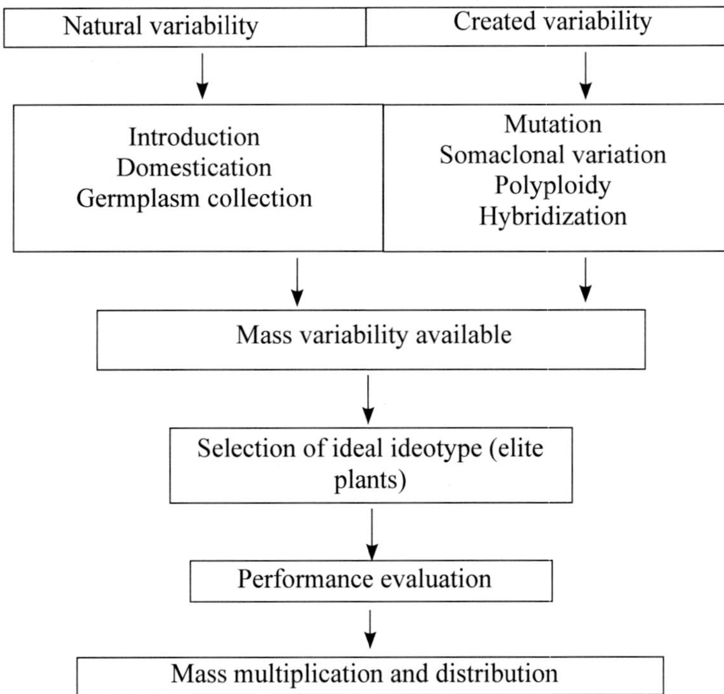

1.8 Plantation Crop Breeding

It is very difficult to define the plantation crops. Opinion varies person to person in classification of plantation crops but generally accepted one is that plantation crops are manmade stand of woody perennial crops which are grown on large contiguous area and require large area and intense management, input, and care and process. There are diverse types of plantation crops like coconut, areca nut, cashew nut, oil palm, rubber, betel vine, palmyrah palm etc. Among them, coconut, areca nut, cashew nut, black pepper and cardamom are small holder plantation crops whereas tea, coffee and rubber are estate crops.

India enjoys first place in coconut, arecanut, cashewnut and tea; second place in cardamom and 4th place in coffee. Kerala is legendary state in plantation crops having more than 50% area of plantation crops and 40% production of coconut. Karnataka is the major state for growing arecanut.

India, Indonesia, Malaysia, Philippines, Mexico, Brazil, Colombia, Guatemala, Sri Lanka etc. are the major countries growing the plantation crops with major crops like coconut, cashew nut, rubber and coffee.

1.8.1. Research Set Up

1916: Four Research Stations were established for coconut at South Kanara, Pilicode, Nileswar and Kasaragod.

1924: Coconut exotic collections were introduced into India from the Philippines, Malaysia, Fiji, Indonesia and Vietnam and were planted in Nileswar and Kasaragod. J.S. Patel of Department of Agriculture under Madras Presidency started genetic improvement work in coconut along with Marshal. They released first coconut variety tall × dwarf at Nileswar during 1932.

1952: For cashew breeding a centre was established at Kottarakkare, Kerala. Later other four centres were established at Ullal, Bapatla, Daragaon and Vengurla. Then for arecanut, CPCRI was established.

In 1600 AD, coffee was introduced by Muslim pilgrim Baba Buadan into India from Yemen, commonly called as Mecca coffee to the Chikkamagalur district in Karnataka.

1920: Commercial plantation of coffee was initiated by British Government through Coffee Board.

Tea: English botanist Sir Joseph Banks suggested East India Company to go for large scale cultivation of tea. Robert Bruce was the first person who identified tea growing in North-Eastern India. Dr.

Chrusti conducted first experiment on tea in Nilgiris/Coimbatore, Tamil Nadu during1830. Indian Tea Committee (ITC) released tea to Coorg (Karnataka), Nilgiris (Tamil Nadu) and Chennai which led the commercial cultivation of tea in South India.

Today CPCRI has recorded world's largest collection of germplasm on plantation crops — coconut 132, arecanut 99, cocoa 134 and oil palm 71.

Review Questions

Part A

Answer the following questions true or false.

1. India is an important centre of origin and diversity of mango. True/False
2. Majority of the fruit species are homozygous. True/False
3. Rootstock should be free from suckering habit. True/False
4. Dwarf growth stature is desirable for scion. True/False
5. Most of the fruit species are polyploid in nature. True/False

Part B

Answer the following questions.

1. Who introduced the grape cultivar 'Anad-e-Shahi' from the Middle East?
2. Origin of fruit culture can be traced back to — years.
3. Define plantation crops.
4. India enjoys first/second place in coconut, arecanut, cashewnut and tea.
5. IIHR stands for —.

Part C

Write a brief note on each of the following.

1. Problems of fruit breeding.
2. Objectives of fruit breeding
3. Some genetic changes occurred in a number of fruit crops during domestication. Substantiate it.

2

Breeding Strategies

2.1 Introduction

The different breeding strategies (methods) that can be employed for improvement of fruit crops include introduction, open pollinated selection, clonal selection, hybrid breeding, polyploidy breeding, mutation breeding, molecular breeding, in vitro breeding and transgenic breeding. Combining traditional approaches with biotechnological approaches will speed up the rate of genetic improvement. Thus, what is required is to adopt a coordinated approach involving breeders, biotechnologists, horticulturists, pathologists, entomologists and statistician for making the cultivation of fruit crops more scientific and more profitable and thereby contributing more to improving the health of the people in a sustainable manner.

2.2 Breeding Systems in Some Fruit Crops

(a) **Predominately self-pollinated**: Apricot, citrus

(b) **Predominately cross-pollinated**: Apple, almond, banana, blue berry, cherry, date, fig, filbert, grapes, mango, olive, papaya, pear, pecan, pistachio, plum, rasp berry, strawberry, walnut.

2.3 Breeding Objectives

The objectives are to improve yield and quality of fruits and nuts. These objectives can be accomplished through genetic manipulation or environmental manipulation or through both. Horticulturists are engaged in developing package of practices and developing techniques of pruning, training for improving yield and quality and further developing post-harvest management practices for storing and increasing shelf-life of fruits. Breeders are engaged in genetic manipulation aiming at improving yield and quality and overcoming post-harvest physiological disorders. The characters to be improved include tree characteristics, fruit characteristics, nut characteristics, resistance to biotic and abiotic stresses etc. The objectives in fruit crops can be met by improving either the scion cultivars or rootstocks or both.

2.4 Breeding Methods

The different breeding methods employed in annual food crops can be applied with modifications because of the differences that fruit crops show. Fruit crops are mostly perennial, highly heterozygous and out-bred (cross-pollinated) in nature. They can be multiplied through both seeds and asexually. Perennial fruit crops have the advantage that once a promising line is selected, it can be asexually propagated (i.e., the genotype can be fixed). Again, a fruit crop cultivar consists of two genotypes — scion genotype and rootstock genotype. Thus, the improvement in fruit crops can be achieved through improvement of either scion genotype or rootstock genotype or both. The standard methods of handling segregating generation such as pedigree, bulk, single seed descent or dihaploidy cannot be directly employed. As plants are heterozygous, F_1s show a lot of variation unlike in annual self-pollinated crops. Hence, selection can be practised in F_1 generation itself without waiting till F_2 generation as in self-pollinated crops. Once a promising tree is found, it can be fixed and multiplied and maintained by vegetative propagation. If the juvenile period is short, one can raise F_1 trees from the seeds and evaluate it and select the best one. Then, the selected best tree can be multiplied through vegetative propagation. If the juvenile period is long, raise seedlings from the seeds and graft it and then evaluated and select the most promising genotype. Again, once the best genotype is identified, it can be propagated vegetatively.

In case of crops that are vegetatively propagated through either budding or grafting, the selected plants are used as source of scions or buds and grafted/ budded onto selected rootstock(s) and replicated trial is conducted in the following generations with continuous increase in the number of replications and sites of testing with the increase in planting materials. Here, one can use micropropagation technique to multiply the selected plant (s) in order to conduct multilocation replicated trial right in the F_2/ F_3 generation (early generations).

In case of seed propagated fruits (e.g., papaya) the standard method of handling F_2 population such as bulk, pedigree, single seed descent or dihaploidy is applied.

2.4.1. Introduction

Plant introduction is taking a genotype or a group of genotypes to a new area where they are not grown before. Plant introduction is coordinated by National Bureau of Plant Genetic Resources (NBPGR), New Delhi. The different steps of introduction are procurement, cataloguing, evaluation, acclimatization and multiplication and distribution. Some examples of introduction of fruit crops are as under:

Banana: Cultivar Lady Finger from Australia as resistant/ tolerant to bunchy top disease, Grand Naine from France.

Citrus: Torocco Blood from USA, Sunramon from Peru, Vainiglia Sanguigno (pigmented) from USA, Tha and Thai-1 from Thailand.

Grapes: Thompson Seedless, Perlette and Beauty Seedless from USA, Kishmish Beli and Kishmish Charni from USSR, Totlocha from Brazil, Dogridge from USA.

Guava: Beaumont G-35 and Indonesia Seedless from Australia, Verdie and M25988 from USA

Mango: Tomy, Zilets, Haden, Sensation, Julie from USA.

Papaya: P141436, Soniyimma, Malinchly from Nigeria, Sun rise, Wilder Sunset from USA.

2.4.2 Open Pollinated Selection

Most of the present-day cultivars of fruits are the result of open pollinated seedling selection from the mixed population. For the selection there must be:

a) Large population

b) Variation must be present in the population

c) Variation must be heritable

Achievements

Mango: Langra, S.B.Chausa, Sukul, Dashehari, Bombay Green

Guava : L-49, Arka Mridula, Allahabad Safeda, Pant Prabhat, Lalit

Litchi : Swarna Roopa

Aonla : Kanchan (NA4), Krishna (NA5), NA6, NA7.

Papaya: CO-1, CO-2, CO-5, CO-6, Pusa Delicious, Pusa Majesty, Pusa Giant, Pusa Dwarf, Coorg Honey Dew

Kagzi Lime: Vikram, Pramalini, Saisarbati, Jaidevi

Lemon: Pant Lemon

2.4.3 Clonal Selection

A clone is a group of plants produced exclusively from a single individual plant through asexual reproduction. Most of the fruit plants are propagated asexually which consists of large number of clones. So, these plants are known as a group of plants derived from a single plant by vegetative means. In other words, all the vegetative progenies of a single plant make a clone.

Selection within seedling populations from higher yielding orchards or from selected elite trees is a method for genetic improvement of the crop. Clonal selection and vegetative propagation is the best option for genetic improvement of heterozygous species. Scion selections are then grafted on to seedling rootstock.

Achievements

Mango: Clone No.51 from Dashehari, MA-1 from Alphonso, Tomy Atkin from Haden, Pusa Surya from Elden

Grapes: Pusa Seedless from Thompson Seedless

2.4.4. Population Improvement

Reciprocal recurrent selection is being applied for improving coffee, cacao, oil palm, rubber and eucalyptus.

2.4.5. Hybrid Breeding

2.4.5.1 Intraspecific Hybridization

Intraspecific hybridization involves crossing between two cultivars from the same species. This type of hybridization followed by selection in the segregating generation will lead to selection of improved genotype or cultivar if one of the two parents is a commercially grown cultivar. Here the two parents have similar ploidy level.

Mango breeders have crossed two different cultivars and selected trees from this cross and labeled them as hybrids. But we know that the fruit crops are highly heterozygous and so the cross between cultivars cannot be said to be F_1 derived from two pure breeding lines/inbreds. Again, F_1 in case of fruit crops is a population of genotypes.

Innovative Method of Developing Hybrids (Roy, 2019)

Instead of just selecting two cultivars, crossing them and releasing the selected tree from the progeny as hybrid, it would be much better if we do the selfing/ inbreeding for one generation and thus obtain partial inbreds. Practice selection in this generation by way of raising seedlings from the harvested seeds and grafting/budding them onto rootstock. Selected partial inbred lines are then crossed to produce the hybrids. The heterozygosity is thus restored and is at a higher level than the original level of heterozygosity. Besides, the selected plant will now contain fewer decreasing alleles and the probability of getting a novel hybrid increases.

Hybrid Development in other Fruit Crop Species

In case of papaya, hybrid is developed by crossing female from dioecious or gynodioecious plant and male from dioecious or androdioecious plant. In this condition, there is no need for emasculation and hand pollination. The sex ratio in the offspring will be 1:1 (female : male).

2.4.5.2 Interspecific Hybridization

Interspecific hybridization leads to sterility in F_1. In case of apple, which is a complex hybrid involving a number of species, crossing programme for improvement can start with two parent crosses and selections in each segregating generation be crossed with a different species. Interspecific hybridization is being carried out in fruit crops such as apple, banana, strawberry etc. in order to introgress gene for resistance to biotic and abiotic stresses. Interspecific hybridization is very widely practiced in some of the fruit crops like citrus, wherein intervarietal hybridization work is very limited as the desired traits are available across the species and there is no barrier. In the development of better-quality fruits, it is used in custard apple breeding programme.

Achievements

Mango: Mallika, Amrapalli, Pusa Arunima, Arka Anmol, Arka Puneet, Arka Aruna, Arka Neelkiran, Ratna, Sindhu, PKM-1, PKM-2.

Guava: Arka Amulya, Safed Jam, Kohir Safed

Litchi: Sabour Madhu, Sabour Priya

Papaya: CO-2, CO-3

2.4.6 *Polyploidy Breeding*

In many fruits and other vegetatively propagated crops, there are cultivars that are diploids, triploids, tetraploids and hexaploids. So, it would be worthwhile to synthesize trees with higher ploidy levels where no polyploidy exists and assess their usefulness. Polyploidy is one of the ways of creating variability in the germplasm. Where the fruit crop is itself a polyploid, it would be better to resynthesize that polyploid afresh from their diploid parents and assess. Heterozygosity is increased by polyploidy. Higher chromosome number seems to promote frequency of recombination. Production of triploid by hybridizing tetraploids with diploids may be useful in obtaining seedless cultivars.

2.4.7 *Mutation Breeding*

Treating a biological material with a mutagen to induce mutation is called mutagenesis. When mutations are induced for crop improvement, the entire operation of induction and isolation of mutants is known as mutation breeding. When there exists little or no variability in a particular trait, one can adopt this breeding technique to create variability for improving that trait. This technique is more useful in vegetatively propagated fruit crops because our objectives are in most cases to modify the existing cultivars for one simple trait. The material for mutagenic treatment could be seed (in case of papaya, cashew, coconut, date palm, passion fruit, phalsa, tamarind and loquat) or meristematic bud in case of vegetatively propagated fruit crops. Between physical (X-rays, gamma rays) and chemical (EMS, MMS) mutagens, the former is more effective in generating mutants. Treatment of seed, whole plant and meristematic tissue such as apical bud with mutagens will lead to development of chimeras which will take a number of generations to remove before one can isolate a solid mutant. In case of adventitious bud technique, a tissue culture technique, the explant is irradiated and adventitious bud is induced which generates adventitious shoot. In this way, non-chimeric mutants (solid mutants) can be obtained.

2.4.7.1 Steps in Mutation Breeding

The various steps involved in mutation breeding are:

- Objective of programme: Objective should be clear cut and well defined.
- Selection of cultivar for mutagen treatment: Locally accepted best cultivar in which improvement is needed.
- Part of plant to be treated: In sexually propagated fruits, seed is treated. In case of asexually propagated fruits, buds or cuttings are used for mutagenic treatment.
- Dose of mutagen- LD_{50}: It is that dose of mutagen which will kill 50 % of the treated individuals.
- Mutagen treatment: Selected plant part is exposed to the desired mutagen dose. In case of irradiation, the plant parts are immediately planted to raise M_1 plants from them. In case of seed treatment, seeds are pre-soaked few hours to initiate metabolic activities and then exposed to mutagen. Treated seeds are sown immediately in field to raise M_1 generation.

Through selfing or clonal propagation M_2, M_3, M_4 etc. generations are derived subsequently.

- Handling of mutagen treated populations.

Achievements

Mango: Rosica from Peruvian cultivar Rosadodelca.

Papaya: Pusa Nanha from local type.

Grape: Marvel seedless from Delight.

Orange: Washington Navel.

Grapefruits: Marsh and Thompson.

2.4.8 *Marker Assisted Selection (MAS)*

Marker Assisted Selection (MAS) has several advantages over phenotypic selection such as decreased number of breeding generations, availability of a uniform method of scoring, no need to use phenotypic scoring until the end and finally, the possibility of obtaining information on the percentage of genome contributed by each parent in the offspring. MAS will be effective with the identification of more molecular markers. Again, unlike annual food crops, only specific qualitative trait needs to be altered in most fruit crops. So, MAS will become popular as more and more allele specific markers become available. For example, self-incompatibility and male sterility are two traits which limit apricot improvement and so molecular markers may be used for early determination of these phenotypes.

2.4.9 *Backcrossing Using Molecular Marker*

In case of seed propagated fruit crops, one can employ backcross breeding method to transfer gene (s) for a particular trait from other cultivar or species. One can take help of molecular marker technology and come up with desired genotype right in the BC_3 generation in comparison to BC_6 in the conventional backcross breeding method.

2.4.10 *In-Vitro Breeding*

There are many applications of tissue culture (*in vitro* culture) in fruit crops. Tissue culture system can be used for quick multiplication of a genotype. Another application involves generation of variability using somaclonal variation or through treatment of cell with specific chemical mutagen in order to screen for resistance to biotic and abiotic stresses. Development of haploid/di-haploid through anther/pollen culture is another application that has usefulness in fruit crops. This will help in rapid generation of homozygous lines which can be used in the development of hybrids. Embryo culture (in the technique called embryo rescue) is used where few viable seeds are obtained from crosses involving edible cultivars (as in case of banana) or in the distant hybridization programme involving different species of a genus. Embryo rescue technique is already in wide spread use for the development of seedless grape. Embryo culture for rescuing hybrids is also used in papaya which would otherwise abort or fail to develop in early fruit developing selections.

2.4.11 Transgenic Breeding

The advent of recombinant DNA technology (genetic engineering) has opened tremendous possibilities for transforming almost any plant by transferring any gene from any organism across taxonomic barriers. The recombinant DNA technology involves the following major steps:

- Isolation of gene of desired plant.
- Insertion of the isolated gene in a suitable vector (making of a recombinant vector).
- Transformation: Insertion of the recombinant vector into a suitable host (organism/cell).
- Selection of the transformed host.
- Multiplication followed by expression of the introduced gene into the host.

Transgenic plants are reported in number of fruit crops, e.g., kiwi fruit, papaya, pecan nut, lime, rough lemon, mandarin, strawberry, walnut, apple, banana, passion fruit, avocado, sweet orange, grape, peach, plum, apricot, pear, almond, trifoliate orange, cranberry, red raspberry, blackberry etc.

Review Questions

Part A

Answer the following questions true or false.

1. The different breeding methods employed in annual food crops can be applied in fruit crops with modifications. True/False
2. Chemical mutagens are more effective in generating mutants. True/False
3. Transgenic plants are reported in number of fruit crops. True/False

4. There are few applications of *in vitro* culture in fruit crops. True/False
5. F_1 in case of fruit crops is a population of genotypes. True/False

Part B

Answer the following questions.

1. Why mutation breeding is more useful in vegetatively propagated fruit crops?
2. What are the advantages of marker-assisted selection?
3. Cite two fruit crops where transgenic plants have been reported.
4. Self-incompatibility and — are two traits which limit apricot improvement.
5. Recombinant DNA technology is also known as —.

Part C

Write a brief note on each of the following.

1. Mutation breeding.
2. In-vitro breeding.
3. Polyploidy breeding.
4. Clonal selection.
5. Marker assisted selection.

SECTION II

BREEDING OF FRUIT CROPS

3

Breeding of Mango

3.1 Introduction

The Mango (*Mangifera indica* L.) belonging to the family Anacardiaceae in order Sapindales, is amongst the most important tropical fruits of the world. In India, it is also known as the 'king of fruits' owing to the delicious quality of the fruit rich in vitamins and minerals. Its long period of domestication is evident from its mention in the ancient Indian scriptures. In India, mango has been valued not only for religious consideration but also for its importance in the economic and cultural life of the society. Mango is rich in Vitamin A and C, flavonoids, carotenes, glucosides, sterols, terpenes, aromatic acids, essential oils, fatty acids and phenolics. It is a powerful nutritive fruit, containing most of the essential substances needed by the human body. The most important single factor limiting the production of mango fruit at the present time is the erratic fruiting habits of this species. Some cultivars produce a good crop in most years and a poor crop in other years, while still other cultivars produce only a few fruits each year. For many years new cultivars have been selected and grown on the basis of size, appearance, and dessert quality with very little attention given to their production capabilities. The improvement of mango needs to explore new recombinants primarily by means of exploiting the breeding methodologies. Diversity or heterogeneity is the main character desired for breeding either natural or manmade. It is required to have vast genetic pool to get new combinations of desired nature and developing new hybrids. It has been mentioned earlier that the development of mango in India is the result of selection from the amateur gardeners. Now, the scientists have developed certain hybrids of mango.

The opportunity for crop improvement in mango is significant and challenging. There is a lot of cultivars and hybrids available in mango but certain intrinsic constraints are involved like long juvenility, high heterozygosity, one seed per fruit, recalcitrant seeds, polyembryony, self-incompatibility and large area requirement for assessment of hybrids. On the other hand, wide range of diversity and ease of vegetative hybrid propagation are the advantages for the breeders. Mango development is mostly based on the selection of clones / chance seedlings. These selections were made for fruit quality only. Seedling selection from known mother plants is another way of selection for better lines. Modern age requirements of a good cultivar are: dwarfness, precocity, profuse and regular bearing, attractive, good sized and quality fruit, absence of physiological disorders, disease and pest resistance, and improved shelf life etc. As far as the improvement of the rootstock is concerned, the main features desired are polyembryony, dwarfness, tolerance to adverse soil and climatic conditions, and scion compatibility.

3.2 Origin and Distribution

Based on the findings of Mukherjee (1997), the centre of origin and diversity of the genus *Mangifera* is established in Southeast Asia. However, the origin of *Mangifera indica* has been a matter of speculation for many years. Mango is originated in the South East Asian or Indo-Burma Region having 41 recognised species.

The cultivated mango cultivars are the result of constant selection by man from original wild plants for over 4000 years. The wild progenies are still available in India in two species, *Mangifera indica* and *Mangifera sylvatica*, which have small fruits with a big stone, thin acidic flesh and long fibres. The knowledge of vegetative propagation gained in the sixteenth century made it possible to produce a large number of cultivars which were far superior to the wild forms. The fruits have little fibre, and are sweet in taste with more flesh.

Mango has been cultivated for thousands of years in India. Its cultivation is found in many countries of South East Asia (the Philippines, Indonesia, Java, Thailand, Burma, Malaysia and Sri Lanka). Introduction of the mango to East and West Africa and subsequently to Brazil is said to have occurred in the sixteenth century. Mexico acquired the mango in the nineteenth century and it entered Florida in 1833.

The mango has been cultivated since ancient times. Buddhist monks are believed to have taken the mango on voyages to Malayasia and eastern Asia in the 4th and 5th centuries B.C. The Persians are said to have carried it to East Africa about the 10th century A.D. It was commonly grown in the East Indies before the earliest visits of the Portuguese who apparently introduced it to West Africa early in the 16th century and also into Brazil. After becoming established in Brazil, the mango was carried to the West Indies, being first planted in Barbados about 1742 and later in the Dominican Republic. It reached Jamaica about 1782 and, early in the 19th century, reached Mexico from the Philippines and the West Indies (National Mango Database).

3.3 Taxonomy and Classification

According to the phylogenetic system of classification of Hutchinson (1959), mango is classified as:

Phylum - Angiospermae

Sub-phylum -Dicotyledones

Division - Lignosae

Order - Sapindales

Family- Anacardiaceae

Genus - *Mangifera* L.

Species - *indica*

Binomial name - *Mangifera indica* L.

The genus *Mangifera* belongs to the order Sapindales in the family Anacardiaceae which is a family of mainly tropical species with 73 genera, with a few representatives in temperate regions. The genus *Mangifera* consists of 69 species and mostly restricted to tropical Asia. The highest diversity occurs in Malaysia, particularly in peninsular Malaya, Borneo and Sumatra representing heart of the distribution range of the genus. The natural occurrence of all the *Mangifera* species extends as far north as 27° latitude and as far east as the Caroline Islands. Wild mangoes occur in India, Sri Lanka, Bangladesh, Myanmar, Sikkim, Thailand, Kampuchea, Vietnam, Laos, southern China, Malaysia, Singapore, Indonesia, Brunei, the Philippines, Papua New Guinea, and the Solomon and Caroline Islands. Maximum species diversity exists in western Malesia and about 28 species are found in this region. *Mangifera* species are mostly distributed below 300 m but can occur at 600-1900 m above sea level. According to Mukherjee (1953), mango has been cultivated for the past 4000 years at least, with over 1000 cultivars under cultivation during this time. *M. sylvatica* and wild *M. indica* can be found in Sikkim and southern China, at altitudes of 600-1900 m above sea level.

Almost all the commercial cultivars of mango are related to a single species *Mangifera indica*. However, a few commercial cultivars of South East Asia belong to other edible species such as *M. altissima. M. caesia, M. cochinchinensis, M. foetida, M. griffithi, M. langinifera, M. longipes, M. macrocarpa, M. odorata, M. pajang, M. pentandra, M. sylvatica* and *M. zeylanica*. There are different reports regarding the number of species in Genus *Mangifera*. Singh (1969) reported 62 species. There are five species of

Mangifera reported from India e.g. *M. andamanica, M. indica, M. khasiana, M. sylvatica* and *M. comptosperma* (Mukherjee,1953).

The five species, *M. sylvatica, M. khasiana, M. andamanica, M. indica* and *M. comptosperma* reported to be found in India, are distinctly different from each other. *M. indica* is the most important species of the genus, as a producer of the most delicious tropical fruit, the mango. It is closely related to *M. longipes* and *M. sylvatica.*

3.4 Morphology

3.4.1 Habit

Mango trees are erect and fast growing with their canopy broad and rounded, or more upright, with a relatively slender crown. The mango tree is medium to large (10 to 40 m) in height, evergreen with symmetrical, rounded canopy ranging from low and dense to upright and open. It attains great age, can live well over 100 years. Trees in cultivated orchards are kept at 6 – 9 m height.

3.4.2 Root

The tree forms a long unbranched taproot (up to 6 to 8 metres and more) plus a dense mass of superficial feeder roots. Feeder roots develop at the base of the trunk or slightly deeper which form anchor roots and sometimes a collection of feeder roots develops above the water table. The fibrous root system extends away from the drip line. Effective root system of an 18-year-old mango tree may observe a 1.2 m depth with lateral spread as far as 7.5 m. In deep soil the taproot descends to a depth of 20 ft, and the profuse, wide-spreading feeder roots also send down many anchor roots which penetrate for several feet.

3.4.3 Stem

Erect, woody, hard, with resinous bark. Bark is usually dark grey-brown to black, rather smooth, superficially cracked or inconspicuously fissured, peeling off in irregular, rather thick pieces.

3.4.4 Leaf

Mango leaves in general are dark green above and pale below. The leaves are petiolate, pulvinus, exstipulate, alternate in phyllotaxy, simple, dorsiventral (also called bilateral, bifacial), leathery, oblong-lanceolate to linear. Leaves are variable in shapes like oval-lanceolate, lanceolate, oblong, linear-oblong, ovate, obovate-lanceolate or roundish-oblong depending on variety. The upper surface is shining and dark green while the lower one is glabrous and light green. The midrib is pale and conspicuous with many prominent light-coloured horizontal veins. One or two growth flushes appear per year and they are located sporadically across the canopy of a given tree. As the leaf grows its colour changes from tan-red to green, passing through many different shades until it becomes shiny dark green above, lighter below, with yellow or white venation at maturity. Leaves may persist several years. The leaves have fibres and when crushed they emit strongly the smell of turpentine. The leaves contain considerable amounts of mangiferin (xanthone).

3.4.5 Inflorescence

The inflorescence is branched panicle borne at shoot terminals, 6.4 to 40.6 cm, possessing many very small (4 mm) greenish white or pinkish flowers. The inflorescence is a narrowly to broadly conical panicle depending upon cultivar and environmental conditions during its development. The colour of the panicle may be yellowish-green, light green with crimson patches or with crimson flush on branches. The branching of the inflorescence is usually tertiary, rarely quaternary, but the ultimate branching is always cymose. Each panicle bears 500 to 6,000 flowers of which 1 to 70 percent are bisexual, the remainder are male depending on the cultivar and temperature during its development. Both male and bisexual flowers are borne on the same tree. The flowers are radially symmetrical, and usually have 5 petals, streaked with red. There is usually only 1 fertile stamen per flower; the 4 other stamens are sterile. The flower has a conspicuous 5-lobed disk between the petals and stamens.

3.4.6 Flower

Mature terminal branches bear pyramidal flower panicles that have several hundred white flowers that are about a 0.6 cm wide when open. Most of the flowers function as males providing pollen, but some are bisexual and set fruit. Pollination is by flies, wasps, and bees. Hundreds and even as many as 3,000 to 4,000 small, yellowish or reddish flowers, 25% to 98% male, the rest hermaphroditic, are borne in profuse, showy, erect, pyramidal, branched clusters of 6-40 cm high. The size of both male and hermaphrodite flowers varies from 6 to 8 mm in diameter. These flowers emit a volatile substance, causing allergic and respiratory problems for some persons. Few of the flowers in each inflorescence are perfect, so most do not produce pollen and are incapable of producing fruit. Panicles that arise later in the bloom season or in shaded parts of the canopy tend to have more hermaphroditic flowers. Panicles are initiated in terminal buds 1-3 months prior to flowering, triggered by low temperatures or seasonally dry conditions.

3.4.7 Calyx

The calyx is usually five partite. The lobes are ovate-oblong and concave.

3.4.8 Corolla

The flowers have four to five petals that are oblong to ovoid to lanceolate and also thinly pubescent. The floral disc is four to five lobed, fleshy and large, and located above the base of the petals. On fading, the petals become pinkish. Between the corolla and androecium there is an annular, fleshy, and five-lobed disc.

3.4.9 Androecium

The androecium consists of stamens and staminodes, altogether five in number, of which usually one, or rarely two, are fertile and the rest are sterile. Although there are four to five stamens, only one or two of them are fertile; the remaining are sterile staminodes that are surmounted by a small gland. All the stamens are inserted on the inner margin of the disc. The position of the fertile stamen and pistil may be either parallel or oblique to each other. In addition, two to three smaller filaments arise from the lobes of the nectaries. The stamens are central. The fertile stamens are longer than the staminodes and are nearly equal to the length of the pistil. The colour of the anther is pink, which turns purple at the time of shedding. The pollen grains are of variable shapes, with the size varying from 20 to 35μm. Small amount of pollen is produced in *M. indica.*

3.4.10 Gynoecium

Sessile, oblique, unilocular (one-celled) ovary set on a disc with lateral style and small simple stigma, approximately same length as fertile stamen, with single anatropous ovule.

3.4.11 Fruit

The fruit is more or less compressed, fleshy drupe. It varies considerably in size, shape, colour, presence of fibre, flavour, taste and several other characters. The most characteristic feature of the mango fruit is the formation of a small conical projection developing laterally at the distal end of the fruit, known as the beak. The fruit ranges from 6.25-25 cm in length and from a few grams to 180-225 g. The fleshy drupe, contains edible mesocarp of varying thickness. It is resinous and highly variable with respect to shape and size. Chlorophyll, carotenes, anthocyanins and xanthophylls are all present in the fruit, although chlorophyll disappears during ripening, whereas anthocyanins and carotenoids increase with maturity. The exocarp (epicarp) is thick and glandular. The mesocarp can be fibrous or fibre-free with flavour ranging from turpentine to sweet. The endocarp is woody, thick and fibrous. This flesh is rich in vitamins A, C and D. The mango flesh is sometimes astringent (turpentine-like), and can have fibres extending from the endocarp (stone). The acrid juice, with turpentine like smell, present in the stalk or sometimes in the fruits, is due to myrcene and ocimene.

3.4.12 Seed

The seed of mango is solitary, large and flat, ovoid oblong, and is surrounded by the fibrous endocarp at maturity. The testa and tegmen are represented as two thin and papery layers. The seed is exalbuminous (non-endospermic) with two fleshy cotyledons. The seeds are not labyrinthine. Seeds of monoembryonic mango types contain a single zygotic embryo, whose cotyledons can be unequal and lobed. The seeds of polyembryonic mango cultivars contain one or more embryos derived directly from the nucellus, a material tissue. Nucellar embryos apparently lack a suspensor. Polyembryony has also been reported in *M. casturi, M. laurina and M. odorata.* Occasionally, a single embryo is found within a seed of a polyembryonic cultivar; however, it may or may not be a zygotic embryo. Certain polyembryonic cultivars reportedly can produce seeds with adventitous nucellar embryos only, e.g. 'Strawberry', 'Carabao' and 'Pico' and 'Olour' and 'Cambodiana'. It is believed that polyembryony is a polygenic trait and segregates as a recessive character in the progeny of controlled crosses. Seedlings can be distinguished whether they are from the single zygotic seeds or from polyembryonic seeds by means of isozyme analysis. Mango seeds are considered to be recalcitrant, and cannot survive for more than a few days or weeks in storage at ambient temperatures. This important characteristic of mango seeds would have prevented their long-distance dispersal until recent times.

3.5. Floral Biology

Mango inflorescence is terminal with frequent emergence of the multiple axillary panicles. Anthesis starts early in the morning and completes at noon. Stigma receptivity remains for 72 hours but most receptive period is for the first 6 hours. Minimum pollen germination time is 1.5 hour. Initial fruit set depends upon the ratio of the perfect to male flowers. Proportion of perfect flowers required for optimum fruit set must not be less than 1%. Flower starts opening early in the morning from 4-7 am and maximum flowers open between 9.30-10.30 am and complete at 11 am. Dehiscence of anthers takes place at 11.30 am and it continues up to 3.45 pm. The pollen grains are oval, or triangular or oblong. Stigma becomes receptive even 18 hours before flower opening. The flowering duration is usually of short i.e. 2 to 3 weeks. The mango inflorescence or panicle bears mainly two types of flowers – male and perfect. The number of flowers per panicle varies between 1000 to 6000 depending upon the variety and climatic factors. The percentage of perfect flowers varies between 0.74 per cent in Rumani, 16.41 to 55.7 per cent in Neelum and up to 69.8 per cent in Langra.

3.6 Pollination

Mangoes are monoecious and self-fertile, so a single tree will produce fruits without cross-pollination. Polyembryonic fruits may not require pollination at all. Branches may be ringed or leaves sprayed with chemicals to induce flowering, but the results are mixed. Mango is self-fertile but cross-pollination increases fruit set. Some self-unfruitful cultivars may get benefit from cross-pollination. Pollen cannot be shed in high humidity or rain. Fertilization is also ineffective when night temperatures are below 22.8°C. There is no air-borne pollen since it is heavy and adherent. Naturally more than 50% flowers do not receive any pollen. Though the ratio of hermaphrodite to male flowers is cultivar related, cool temperatures may also influence sex expression to favour majority of male flowers. There are several hundred flowers in a panicle and less than 1% only develops fruits because of pollination failure and premature fruit drop. The actual degree of self-fertility and sterility in individual cultivars has not been determined, but there is some variation. Though self-sterility is not a major problem in fruit set, but within a cultivar, there is a definite need for a pollinating agent. The cultivars of mango like Dashehari, Langra and Chausa were found to be self-incompatible. The effect of cool weather adversely affects pollen tube growth, but this was not considered to be of major importance. The need for cross-pollination between mango cultivars is not critical, at least for most cultivars, but pollinating insects are needed to pollinate within cultivar to get satisfactory crop. Several agents have been attributed as pollinators of mango. Important pollinators are wind, but insects like honey bees, ants, and house flies play an important role. Honeybees are the most important hymenopterous insect pollinators of mango flowers. Lack of adequate fruit set in large plantings particularly of single cultivars is frequent. The mango flowers do not appear to be attractive to honey bees as they tend to open when many other flowers are also available leading to poor visitation in commercial mango orchards.

3.7 Emasculation and Crossing Techniques

Since a large number of male and perfect flowers are borne on a mango panicle, it requires a special crossing technique. The panicle should be bagged with a muslin bag (60 cm x 30 cm) fully stretched and fixed with two rings and a rod made of spliced bamboo. A piece of thick iron wire can also be made into a good frame for stretching the muslin bag over the panicle. Staminate flowers of the selected panicle to be used as female parent should be removed daily before dehiscence. Panicles of the cultivar selected as male parent should also be bagged before their flowers begin to open. Freshly dehisced male flowers should be carried in a small petridish lined with a filter paper and covered with another petridish to protect the flower from contamination with foreign pollen carried by insects. Perfect flowers should be emasculated early in the morning before dehiscence of anthers. Freshly dehisced anther of the male parent should gently be brushed against the stigma which should then be examined under lens to see if pollen grains have adhered to it. As the pollination of flowers in any one panicle is carried over a number of days, only the pollinated flowers should be allowed to remain on the panicle. It has been found advantageous to keep the panicles enclosed in bags till the fruits set and develop slightly. The conventional method of pollination is time consuming, cost intensive and inefficient because of tallness and difficulty to handle trees and poor fruit set. ' Caging technique' for crossing, developed at IARI following the discovery of self-incompatibility in Dashehari, Langra, Chausa and Bombay Green, involves planting of grafted plants of the self-incompatible cultivar along with those of male parents enclosed in an insect proof cage and allowing pollination by freshly reared house flies and thus doing away with the tedious hand pollination.

3.8 Germination

Germination is hypogeal. Seeds are usually planted with the endocarp (stone) attached and remain viable for about a month after extracting from fully ripe fruits. They are best stored in charcoal. Seeds are sometimes shelled (endocarp removed) and this hastens germination. Germination of seed stones takes approximately 20 days and seedlings reach a height of about 6 inches at 6-8 weeks, and 9-12 inches at 10-12 weeks.

Zygotic embryos in monoembryonic seeds do not breed true. Apomictic embryos in polyembryonic seeds will have the same genotype as the parent; but zygotic seedlings may be present. Occasionally, seedlings with multiple shoots from buds in the axils of the cotyledons are produced, but they will have only one root.

3.9 Genetics and Cytogenetics

Mango (*Mangifera indica* L.) has 40 somatic chromosomes (2n=40, X=10). Mukherjee (1953) suggests it to be an allopolyploid. Genetic studies carried out using over a thousand hybrid progenies by Sharma and Majumder (1988) showed that dwarf stature, regular bearing and precocity are controlled by recessive genes. Regularity in bearing appeared to be linked to precocity, and, biennial bearing is dominant over regular bearing habit. As for skin colour, it was found that when red coloured cultivars were crossed with green coloured cultivars, gradation of colour in the progenies indicated that fruit colour was controlled by a number of loci. Presence of a beak on the fruit seems to be dominant when 'Totapuri' was used as one of the parents. Bunch-bearing was found to be dominant over single-fruiting. Cytoplasmic inheritance was observed for resistance to bacterial canker as all the progenies were found to be susceptible when Neelum was used as the female parent, irrespective of the male parents used. Mango, as a breeding material, is difficult to be handled because of its long juvenile phase and high heterozygosity with only a small number of hybrid progenies recoverable. However, information obtained on inheritance of characters that follows should be of help in planned hybridization (Dinesh *et al.*2016).

1. Upright habit of the tree as dominant over spreading, and, spreading as dominant over dwarfness
2. Dwarfness found to be governed by recessive genes
3. Strong linkage found between bearing and fruit quality
4. Biennial bearing tendency found to be dominant over regular-bearing
5. Regularity of bearing controlled by recessive genes, found to have close linkage with precocity in bearing
6. Bunch-bearing habit observed to be dominant over single-fruit bearing
7. Presence of beak, marked with sinus, dominant
8. Transgressive segregation for fruit size observed in F_1 hybrids
9. Red skin-colour found to be dominant and gradations in skin colour of F_2 suggestive of the role of multiple genes; also, the case with flesh colour
10. Resistance to floral malformation seems to be controlled by recessive genes
11. Cytoplasmic inheritance found in the case of bacterial canker, and
12. Spongy tissue observed to be genetically controlled

3.10 Breeding Objectives

Most of the commercial mango cultivars have a strong biennial bearing tendency. Thus, the main objectives of breeding have been regular bearing and good fruit quality. An ideal mango cultivar should be dwarf in growth habit, attractive (golden apricot) in colour on ripening, medium size (200 g) and good quality fruit (high pulp: stone ratio, firm and fibreless flesh), precocious and regular in bearing, highly tolerant to various diseases including mango malformation and pests like mango hopper, and high keeping quality. A good mango cultivar should have a high ratio of 3.31 to 4.0 of edible to non-edible matters.

3.11 Breeding Methods

Mango breeding studies began in the 17th century in India by selection of seedlings from open-pollinated crosses and propagation of these seedlings through inarching. Since then several breeding programmes have released hundreds of cultivars in different countries. Although difficult to apply, hand pollination allows a better control of parents and progenies facilitating inheritance studies. The low success in obtaining hybrid progeny, the need for large area for seedling evaluation, and the long juvenile period are discouraging aspects of mango breeding. Therefore, the development of new procedures or techniques to improve hand and open-pollination as well as the handling of the population of hybrid progeny are highly essential for efficient mango breeding. Improvement in the hybridization technique, using a few flowers in a larger number of panicles for crossing, and then, covering the panicles with polythene bag instead of the muslin bag, resulted in raising a large number of hybrid progenies. In fact, one of the main reasons for success in later work on hybridization was the use of information on gene donors, and prevention of indiscriminate crossing among cultivars.

3.11.1 Introduction

For incorporation of good colour to boost export of fresh fruits, a number of mango cultivars were introduced from different countries for use as donor parents. Tommy, Ziulete, Haden, Sensation and Julie are the coloured cultivars of mango which were introduced from Miami, Florida (USA). Other cultivars, Pl 24927, M 4336 (Carabao) from USA and EC 201556 (Carabao) from Philippines were introduced as regular bearing cultivars, Cultivar Amolie and Sweet were introduced from Belgium and Thailand respectively.

3.11.2 Selection

Almost all the present commercial cultivars of mango in the world were developed from open-pollinated seedling selection e.g. Dashehari, Langra, S.B.Chausa, Rataul, Swarnarekha etc. The evolution of Florida cultivars which are the leading mango cultivars of the world is interesting. In 1889, introductions were made from India of which Mulgoa became popular. Cultivar Haden was a seedling of Mulgoa. Subsequently, many promising seedlings were selected which became popular. Tommy Atkins from Haden, Keitt from Mulgoa, Dyke and Palmer from unknown origin, Irwin from Lippins, Golden Nuggets and Brooks from Sandersha, Sensation from unknown origin etc. are promising seedling selections.

***Niranjan*:** This is an off-season bearing selection from the cultivar, 'Royal Special'. 'Niranjan' flowers thrice a year (August, October and December). Flowering intensity is highest during December. It has a low flowering intensity in general, and moderate percentage of perfect flowers. Fruits are round, with low pulp content. This cultivar was released by Gujarat Agricultural University.

***Alphonso 900*:** This is a selection from the cultivar, Alphonso. It is an early-bearer with uniform-size fruits of excellent quality, pleasing flavour, good in sugar: acid blend, attractive fruit colour and a long shelf-life, suitable for processing. This cultivar was developed at Konkan Krishi Vidyapeeth, Dapoli.

***Pusa Surya*:** This is a selection from the cultivar, Eldon, released by Indian Agricultural Research Institute, New Delhi. It bears medium-sized fruits, has a red peel colour (similar to that of 'Sensation').

Subhash: This is a selection from seedlings of 'Zardalu'. It is a mid-season cultivar. Ripe fruits are bright yellow as in 'Zardalu', with the shape of 'Langra' fruit. Fruits are medium in size, with an average weight of 220g. TSS and acidity of the fruits are 24°Brix and 0.29%, respectively. Pulp content is 76%.

***Menaka* :**'Menaka' is a selection arising from 'Gulabkhas'seedling. It is a regular-bearing and late-maturing cultivar; fruits are attractive with deep-red basal portion. Pulp is deep yellow, sweet and pleasant in flavour, low in fibre and firm. Fruit shape is oblong-oblique. Average fruit weight is 300g, TSS of the fruit is 20°Brix, acidity 0.14% and pulp content 75%.

3.11.3 Clonal Selection

Exploitation of natural variability for selection of superior clones of commercial mango cultivars has been undertaken. Clonal selection has also resulted in identification of few elite clones. Dashehari-51 from Dashehari, a regular bearer (CISH, Lucknow), ' Subash', a chance seedling from Zardalu (BAC,

Sabour), Red blush, a strain of Alphonso (Vengurla), heavy yielding strains of Langra and Himsagar (Kalyani, W.B.), bacterial black spot resistant clones of Kensington, superior clones of Rumani and Neelum (Tamil Nadu) and a regular bearing cultivar 'Cardoz Mankhurad' in Maharashtra which is selected from Goa Mankurad. In Maharashtra, one off-season selection 'Niranjan" has been made at Parbhani, which flowers during June to July and its fruit matures in October. In TNAU (Regional Research Station, Paiyur), a clonal selection from Neelum was identified as dwarf cultivar and released as Paiyur-1. This is suitable for high density planting (400 plants/ha).

Arka Neelachal Kesari: This is a cultivar released from Central Horticultural Experiment Station, Bhubaneswar. It is a clonal selection from 'Gulabkhas'. Fruits are medium-sized and average fruit weight is 220g per fruit; pulp is light-yellow, sweet to taste and excellent in quality, with a good sugar-acid blend.

Pant Chandra: This is a clonal selection from 'Dashehari' from Govind Ballabh Pant University of Agriculture & Technology, Pantnagar. Plants are tall, with an erect growth habit. Fruit at maturity remains green. It is a mid-season cultivar. Fruits weigh upto 150g. Fruit pulp is reddish-yellow, with total soluble solids at 18%, and having a pleasant aroma.

Pant Sinduri: This is a clonal selection from 'Dashehari' from Govind Ballabh Pant University of Agriculture & Technology, Pantnagar. Trees are medium in height, with a round top canopy. Fruit is yellow, with a pink shoulder. Average fruit weight is 200g. Fruit pulp is yellow, with a pleasant aroma. Total soluble solids vary from 16 to18%, with an average yield of upto 150 kg per tree. Fruits mature from the last week of May to the first week of June.

3.11.4 Hybridization

In mango hybridization, work taken up in post-independence period laid emphasis on regular and precocious bearing, dwarfness, high percentage of pulp, fibreless flesh, large fruits with red blush, good keeping quality and freedom from spongy tissue. Mango breeding studies, using inter-varietal hybridization, have been improved in many countries. The low fruit set and long juvenile period affect evaluation, selection and releasing of hybrid cultivars. Additionally, the needs of large areas to evaluate the hybrid seedlings are expensive to maintain. The development of techniques to improve natural crossing (open-pollination) and artificial crossing (hand pollination) as well as the management of hybrid population are needed. Although hand pollination is more difficult than open-pollination, it allows a better control of parents and progenies. Now, more comprehensive knowledge about the phenology, inheritance patterns of mango and advanced techniques for hybridization is available. Many environmental and physiological factors related to the undesirable character of mango cultivars (irregular bearing, susceptibility to diseases and pests, poor eating and keeping quality, etc.) are closely controlled by genes. To overcome these, plant breeding can play an important role and work should be done in three directions as introduction, selection and hybridization.

3.11.4.1 Interspecific hybridization

Interspecific hybridization did not receive more attention but it can be a useful tool to transfer some useful genes into cultivated varieties. This is possible because all the *Mangifera* species have the same chromosome number (2n=40). Therefore, they can inter-cross easily (Mukherjee, 1963).

3.11.4.2 Improved Hybridization Techniques

i. Single day pollination of limited number of flowers in a panicle is the ideal practice. Here, the main emphasis is given on utilizing large number of panicles and crossing whatever few flowers opened on the panicle during that single day. Bagging with perforated polythene bags of 24" x 12"size of 100 gauges is preferred. Crossing of a few flowers in a given panicle at one time is advocated than taking up crossing in more number of flowers in a given panicle in batches over a number of days (Mukherjee *et al.*,1961).

ii. **Caging technique:** The discovery of self-incompatibility in some of the popular cultivars at IARI, New Delhi led to further improvement in the technique of hybridization. It is known as caging technique (Sharma and Singh, 1970). In this technique, grafted plants of parent cultivars are enclosed in an insect proof cage and pollination is effected through freshly reared houseflies.

iii. **Marker gene:** The purple colour of new leaves and panicle and beak characters of fruit help in identifying the hybrid seedlings in the nursery.

iv. A new off-season crossing technique is suggested by Kulkarni (1986). It involves induction of flowering in the desired parents in off-season by veneer grafting. The defoliated shoots of the desired parents are grafted on to the leafy shoots of off-season flowering cv Royal special, and open-pollination is allowed between the desired parents. As no other cultivar flowers during this season, this is a safe technique.

Important Mango Hybrids:

Mallika: This hybrid between Neelum and Dashehari was released from IARI, New Delhi. It has a strong tendency to regular bearing. Fruits, on average, weigh about 350-400 g and have a deep yellow pulp, high TSS, good odour, uniform fruit size and moderate keeping quality.

Amrapali: This is from the parentage Dashehari × Neelum. Plants are dwarf and have a regular bearing habit. Fruits weigh, on average, about 180-250g being borne on clusters, are sweet to taste and have a good keeping quality.

Ratna: This is a hybrid from the cross, Alphonso × Neelum, released by Fruit Research Station, Vengurla. It is regular bearing, produces medium-sized fruits weighing, on average, about 250g. The pulp is orange in colour and is free from spongy tissue or fibre.

Sindhu: This is a hybrid progeny derived by backcrossing Ratna with Alphonso, released by Fruit Research Station, Vengurla. Fruits are borne in clusters and weigh, on average, about 150-220g. The pulp is deep yellow with good sugar: acid blend. Fruits are almost seedless, with a very thin stone, although fruits weighing above 200g have a well-developed seed.

Konkan Ruchi: This is a hybrid from the parentage, Neelum × Alphonso. It bears medium-sized fruits. Konkan Krishi Vidyapeeth, Dapoli developed this hybrid, especially for pickle-making.

Konkan Raja: This is a hybrid from the parentage, Bangalora × Himayuddin, developed by Konkan Krishi Vidyapeeth, Dapoli. It has a compact growth habit and bears large-sized fruits (616g) in clusters. It is good in taste; immature fruits are useful in salad making. Pulp percentage is relatively high (83%), with good TSS (16.8°Brix). It is a regular bearing hybrid.

Pusa Arunima: This is from the parentage, Amrapali × Sensation, released by Indian Agricultural Research Institute, New Delhi. Fruits are medium-sized, with an attractive skin colour. The pulp is deep yellow with TSS of around 20 °Brix.

Pusa Prathibha: This is a hybrid between the cross, Dashehari × Amrapali, developed by Indian Agricultural Research Institute, New Delhi. It is a regular-bearing hybrid, with an attractive fruit shape, bright-red peel and orange pulp. It has oblong, uniform-sized fruits and good sugar: acid blend. The plants are semi-vigorous.

Pusa Shresht: This is a hybrid between the cross, Amrapali × Sensation, developed by Indian Agricultural Research Institute, New Delhi. Trees are semi-vigorous, regular bearing, with elongated fruits and an attractive red peel. The pulp is orange in colour, fibreless and firm when ripe, has a moderate sugar: acid blend, with uniform fruit-size (228g). It contains good amounts of β-carotene and ascorbic acid.

***Pusa Pitamber*:** This is a hybrid between the cross, Amrapali × Lal Sundari, developed by Indian Agricultural Research Institute, New Delhi. The plants are semi-vigorous, regular bearing, with attractive oblong fruits. Fruits turn a uniform yellow on ripening. It has a good sugar:acid blend and uniform-sized fruits.

***Pusa Lalima*:** This is a hybrid between the cross, Dashehari × Sensation, developed by Indian Agricultural Research Institute, New Delhi. The plants are semi-vigorous, regular bearing, with attractive oblong fruits and bright-red peel on yellowish green background, with an orange pulp and good sugar: acid blend.

Ambika: This is a hybrid between the cross, Amrapali × Janaradhan Pasand, developed by Central Institute for Subtropical Horticulture, Lucknow. The fruits of this hybrid are medium in size, with a slight sinus and beak, and a broadly pointed apex. Peel is smooth and tough. Fruits are bright yellow with a dark-red blush. Pulp is firm, with scanty fibre. TSS of this variety is 21°Brix. It is a late-maturing hybrid.

***Arunika*:** This is a hybrid between the cross, Amrapali × Vanraj, developed by Central Institute for Subtropical Horticulture, Lucknow. Fruits of this hybrid are attractive, and with a red-blush, high TSS (24°Brix) and high in carotenoids. Pulp is firm. It is a regular bearer and plants are dwarf.

***PKM-1*:** This is a hybrid released from Horticultural Research Station, Periyakulam. It is of the parentage, Chinnaswarnarekha × Neelum. It is regular in bearing, produces good quality fruits in clusters.

***PKM-2*:** This is a hybrid released from Horticultural Research Station, Periyakulam. It is from the parentage, Neelum × Alphonso. It is regular in bearing and produces good quality fruits in clusters.

***Neeleshan Gujarat*:** This is a hybrid from the parentage, Neelum × Baneshan, developed by Agricultural Experimental Station, NAU, Paria, Gujarat. It is a mid-season variety maturing in May. It is semi-vigorous and is a poor yielder.

***Neeleswari*:** This is from the parentage, Neelum × Dashehari, developed by Agricultural Experimental Station, NAU, Paria, Gujarat. It is a mid-season variety maturing in May. It is vigorous and a moderate yielder. Fruits are medium in size and good in quality, with high pulp content. In appearance, the fruits are similar to 'Langra'.

***Neelgoa*:** This is a hybrid from the cross, Neelum × Mulgoa, developed at Fruit Research Station, Sangareddy, Andhra Pradesh. It has high flowering-intensity and percent perfect flowers. Shape of the fruit is similar to 'Banganapalli'. It is semi-vigorous and a moderate yielder. Fruits are medium in size, good in quality with a high pulp-content.

***Neelphonso*:** This is from the parentage, Neelum × Alphonso, developed by Agricultural Experimental Station, Navasari, Agricultural University, Paria, Gujarat. It has low flowering intensity. Fruits are large in size and excellent in quality, but with low pulp content.

***Sonpari*:** This is a hybrid from the cross, Alphonso × Baneshan, developed by Agricultural Experimental Station, Navasari, Agricultural University, Paria, Gujarat. Fruits are round and weigh about 550g. The peel attains golden-yellow colour when ripe, with TSS of about 19.3°Brix.

***Au-Rumani*:** This is a hybrid by the combination, Rumani × Mulgoa, released by Fruit Research Station, Kodur. It bears large fruits of good flavour. It is a heavy yielder, with moderately firm pulp.

***KMH 1 (Kodur Mango Hybrid 1)*:** This is a hybrid released by Fruit Research Station, Kodur. It is from the parentage, Cherukurasam × Khader. Plants are semi-dwarf, regular in bearing; fruits are fibreless with high Brix value and low acidity.

***Manjeera*:** This is released from Fruit Research Station, Sangareddy. It is from the parentage, Rumani × Neelum. It produces round fruits, with a firm pulp, and bears regularly. The pulp of the fruit is yellow with TSS of about 19°Brix.

***Swarna Jehangir*:** This is a hybrid from the cross, China Suvarnarekha × Jehangir, developed at Fruit Research Station, Sangareddy. It is a late hybrid, maturing in June. Fruits weigh upto 450g. TSS of the fruit is 19°Brix, acidity 0.58%, and pulp content about 77%.

***Neeluddin*:** This is from the parentage, Neelum × Himayuddin, developed at Fruit Research Station, Sangareddy, Andhra Pradesh. It is a mid-season hybrid maturing in May-June. It has moderate flowering-intensity, with medium percentage of perfect flowers. Fruits are large, with average weight of 435g, with TSS and acidity at 18° Brix and 0.46%, respectively.

***Neeleshan*:** This is developed at Fruit Research Station, Sangareddy, Andhra Pradesh. It is a hybrid between the cross Neelum × Baneshan. It is a mid-season hybrid bearing oval shaped fruits weighing upto 400g, TSS is 18.2°Brix, and the pulp content 72 %.

***Sabri*:** This is a hybrid between the cross, Gulabkhas × Bombay Green, developed at Bihar Agricultural University, Sabour, Bhagalpur. It is semi-vigorous with low yield potential. Fruits are small but good in quality, with TSS of 19-20° Brix.

***Mahmood Bahar*:** This is a hybrid from the cross, Bombay Green × Kalapadi, developed by Bihar Agricultural University, Sabour, Bhagalpur. It is a mid-season hybrid where fruits mature in May. It is semi-vigorous and a moderate yielder. Fruits are medium in size and good to taste.

***Al Fazli*:** This is from the parentage, Alphonso × Fazli, released by Fruit Research Station, Sabour. It is superior to Fazli and does not have spongy tissue. Fruits are sweet to taste.

***Jawahar*:** This is a hybrid between the cross, Gulabkhas × Mahmood Bahar, developed by Bihar Agriculture University, Sabour (Bihar). It is a mid-season hybrid. Fruits are medium in size with average weight of 215g. The pulp is light-yellow, sweet to taste, pleasant in flavour and remains firm even after ripening. TSS, acidity and pulp percentage are 22.5°Brix, 0.14% and 79.5%, respectively.

***Sundar Langra*:** This is from the cross, Sardar Pasand × Langra. It is regular-bearing and the fruit resembles 'Langra' in shape and size. Fruits are medium-sized and sweet to taste.

***Arka Aruna*:** This is developed at Indian Institute of Horticultural Research, Bangalore. It is from the parentage, Banganapalli × Alphonso. It is regular-bearing; pulp is free from fibre or spongy tissue, pale-yellow in colour, moderately firm, good for making mango bars. Fruit size is large. Plants are dwarf in stature.

***Arka Puneet*:** This is developed at Indian Institute of Horticultural Research, Bangalore. It is from the parentage, Alphonso × Banganapalli. It has an attractive fruit skin colour, medium sized, fruits free from spongy tissue, with good keeping quality and sugar-acid blend.

***Arka Anmol*:** This is developed at Indian Institute of Horticultural Research, Bangalore. It is from the parentage, Alphonso × Janardhan Pasand. It is regular-bearing, has fruits with uniform yellow skin colour. It is free from spongy tissue, has good keeping quality and a good sugar-acid blend.

***Arka Neelkiran*:** This is developed at Indian Institute of Horticultural Research, Bangalore. It is from the parentage, Alphonso × Neelum. It is regular-bearing, with medium-sized fruits free from spongy tissue, good pulp colour, excellent skin colour, and the tree is semi-vigorous.

***Arka Udaya*:** This is from the cross Amrapali × Arka Anmol. Fruits weigh 225 to 250g, and are oval in shape; pulp is deep orange in colour, firm and fibreless. TSS is 21°Brix, and pulp recovery is over 70%. Fruits have an excellent shelf-life. It is a late-season hybrid, with a semi-vigorous growth habit.

3.12 Cultivars Released

They have already been described above. There are nearly 1000 mango cultivars in India. However, only about 20 cultivars are grown commercially. Most of the Indian mango cultivars have specific ecogeographical requirements for optimum growth and fruiting.

Characteristics of important commercial mango cultivars are as under:

1. **Alphonso**: This is the leading commercial cultivar of Maharashtra state and one of the choicest cultivars of the country. This cultivar is known by different names in different regions, *viz.* Badami, Gundu, Khader, Appas, Happus and Kagdi Happus. The fruit of this cultivar is medium in size, ovate oblique in shape and orange yellow in colour. The fruit quality is excellent and keeping quality is good. It has been found good for canning purpose. It is mainly exported as fresh fruit to other countries. It is a mid-season cultivar.

2. **Bangalora**: It is a commercial cultivar of south India. The common synonyms of this cultivar are Totapuri, Kallamai, Thevadiyamuthi, Collector, Sundersha, Burmodilla, Killi Mukku and Gilli Mukku. The fruit size is medium to large, its shape is oblong with necked base and colour is golden yellow. Fruit quality is poor but keeping quality is very good. It is widely used for processing. It is a mid-season cultivar.

3. **Banganapalli**: It is a commercial cultivar of Andhra Pradesh and Tamil Nadu. It is also known as Chapta, Safeda, Baneshan and Chapai. Fruit is large in size and obliquely oval in shape. The colour of the fruit is golden yellow. Fruit quality and keeping quality are good. It is a mid-season cultivar and is good for canning.

4. **Bombai**: It is a commercial cultivar from Bihar state. It is also known as Malda in West Bengal and Bihar. Fruit size is medium, shape ovate oblique and colour is yellow. Fruit quality and keeping quality are medium. It is an early season cultivar.

5. **Bombay Green**: It is commonly grown in north India due to its early fruiting quality. It is also called Malda in northern India. Fruit size is medium, shape ovate oblong and fruit colour is spinach green. Fruit

quality is good and keeping quality is medium. It is a very early season cultivar.

6. **Dashehari**: This cultivar derives its name from the village Dashehari near Lucknow. It is a leading commercial cultivar of north India and one of the best cultivars of our country. The fruit size is small to medium, shape is oblong oblique and fruit colour is yellow. Fruit quality is excellent and keeping quality is good. It is a mid-season cultivar and is mainly used for table purpose.

7. **Fazli**: This cultivar is commonly grown in the states of Uttar Pradesh, Bihar and West Bengal. Fruit is very large, obliquely oval in shape. Fruit colour is light chrome. Fruit quality and keeping quality are medium. This is a late season cultivar.

8. **Fernandin**: This is one of the oldest cultivars of Bombay. Some people think that this cultivar originated in Goa. Fruit size is medium to large, fruit shape is oval to obliquely oval and fruit colour is yellow with a blush of red on shoulders. Fruit quality and keeping quality are medium. It is a late season cultivar mostly used for table purpose.

9. **Himsagar**: This cultivar is indigenous to Bengal. This is one of the choicest cultivars of Bengal and has gained extensive popularity. Fruit is of medium size, ovate to ovate oblique in shape. Fruit colour is yellow. Fruit quality and keeping quality are good. It is an early season cultivar.

10. **Kesar**: This is a leading cultivar of Gujarat with a red blush on the shoulders. Fruit size is medium, shape oblong and keeping quality is good. It is an early season cultivar.

11. **Kishan Bhog**: This cultivar is indigenous to Murshidabad in West Bengal. Fruit size is medium, fruit shape oval oblique and fruit colour is yellow. Fruit quality and keeping quality are good. It is a mid-season cultivar.

12. **Langra**: This cultivar is indigenous to Varanasi area of Uttar Pradesh. It is extensively grown in northern India. Fruit is of medium size, ovate shape and lettuce green colour. Fruit quality is good but keeping quality is medium. It is a mid-season cultivar.

13. **Mankurad**: This cultivar is of commercial importance in Goa and in the neighbouring Ratnagiri district of Maharashtra. The cultivar develops black spots on the skin in rainy season. Fruit is medium in size, ovate in shape and yellow in colour. Fruit quality is very good but keeping quality is poor. It is a mid-season cultivar.

14. **Mulgoa**: This is a commercial cultivar of south India. It is quite popular among the lovers of mango owing to high quality of its fruit. Fruit is large in size, roundish oblique in shape and yellow in colour. Fruit quality and keeping quality are good. It is a late season cultivar.

15. **Neelum**: This is a commercial cultivar indigenous to Tamil Nadu. It is an ideal cultivar for transporting to distant places owing to its high keeping quality. Fruit is medium in size, ovate oblique in shape and saffron yellow in colour. Fruit quality is good and keeping quality is very good. It is a late season cultivar.

16. **Samarbehisht Chausa**: This cultivar originated as a chance seedling in the orchard of a Talukdar of Sandila district Hardoi, U.P. It is commonly grown in northern part of India due to its characteristic flavour and taste. Fruit is large in size, ovate to oval oblique in shape and light yellow in colour. Fruit quality is good and keeping quality is medium. It is a late season cultivar.

17. **Suvernarekha**: This is a commercial cultivar of Visakhapatnam district of Andhra Pradesh. Other synonyms of this cultivar are Sundari, Lal Sundari and Chinna Suvernarekha. Fruit is medium in size and ovate oblong in shape. Colour of the fruit is light cadmium with a blush of jasper red. Fruit quality is medium and keeping quality is good. It is an early season cultivar.

18. **Vanraj** : It is a highly prized cultivar of Vadodra district of Gujarat and fetches good returns. Fruit is medium in size, ovate oblong in shape and colour is deep chrome with a blush of jasper red on the shoulders. Fruit quality and keeping quality are good. It is a mid-season cultivar.

19. **Zardalu** : This cultivar is indigenous to Murshidabad in West Bengal. It derives its name from Zardalu, a dry fruit popular in North West Frontier Province and Sindh in Pakistan owing to similarity in shape. Fruit size is medium, oblong to obliquely oblong and golden yellow in colour. Fruit quality is very good. Keeping quality is medium. It is a mid-season cultivar.

20. **Sai Sugandh:** This is a late cultivar, maturing in June. It is semi vigorous, with fruits showing prominent lenticels and a beak. Fruits are medium in size, of good quality, but susceptible to anthracnose. This cultivar was developed at Konkan Krishi Vidyapeeth, Dapoli.

3.13 Biotechnological Approaches

Use of various biotechnological methods for studying crop genetics and applying it as a tool for fruit crop improvement, is of recent origin. Use of molecular markers, *in vitro* selection and regeneration, identification of linkage for specific traits, and construction of genetic maps, gene pyramiding, allele mining, etc. are some of the recent developments. Discovery of molecular markers has led to detailed genetic analysis and use of newer approaches to improvement in crop plants. Most fruit tree breeding programmes follow the scheme of classical recurrent selection. This results in multiple breeding and production populations. In addition, trees have a long juvenile period. All these lead to a long breeding period. With the advent of marker assisted breeding (MAB), we can plan to reduce this period. The technique has shortened the time required for developing new cultivars. MAB has also made the process more cost-effective than selection based exclusively on the phenotype. However, basic research is needed to understand molecular and physiological mechanisms underlying the trait under study. MAB is based on identification of heritable DNA-sequence differences (polymorphisms). This combines both traditional breeding strategies and molecular tools for selecting plant material with the traits of interest, such as colour, size, or, biotic/abiotic stress resistance. Several types of markers are used in plant breeding, like SSR (simple sequence repeats), AFLP (amplified fragment length polymorphism), SCAR (sequence characterized amplified region), etc. In mango, recently, several groups have made an attempt to develop genomic resources. However, efforts need to be made to develop high-density linkage and association maps that can help breeders identify QTLs and genes responsible for particular traits.

Review Questions

Part A

Answer the following questions true or false.

1. In India, mango is also known as the 'queen of fruits'. True/False
2. 'Alphonso' is also known as 'Badami'. True/False
3. Upright habit of the mango tree is recessive to spreading. True/False
4. Most of the commercial mango cultivars have a regular bearing tendency. True/False
5. Mangoes are monoecious. True/False

Part B

Answer the following questions.

1. Dashehari-51 is selected from ………..
2. 'Mallika' is a hybrid between ----- and ------.
3. What is the centre of origin of *Mangifera indica*?
4. The family Anacardiaceae belongs to order ----------.
5. *Mangifera indica* L. has 20/40 somatic chromosomes.

Part C

Write a brief note on each of the following.

1. Pollination of mango.
2. Improved hybridization techniques of mango.
3. Mango floral biology.
4. Inflorescence and flower of *Mangifera indica.*
5. Origin and distribution of mango.

4

Breeding of Banana

4.1 Introduction

Banana (*Musa* spp.) is one of the oldest fruits and second largest growing fruit crop in the world. It is also known as Adam's Fig and Apple of Paradise. Important banana growing states are Maharashtra, Karnataka, Kerala, Tamil Nadu, Andhra Pradesh, Odisha, Bihar, West Bengal and Assam. In South India, banana is extensively used in all auspicious occasions such as wedding, festival and worship. Banana is a good table fruit. The cultivar Nendran is used for cooking; it is also used for preparation of halwa, sweet, chips etc. Main bud or heart of banana bunches is also used as vegetable in India, Thailand, Philippines, Malaysia and Indonesia. *Musa textiles* is known for strong fibre quality. Ripe banana contains about 20% sugar (but no starch) and about 4.7% protein. Further, it contains vitamins A, B, C, D and E and minerals such as K, P, Ca and Fe.

4.2 Origin and Distribution

The centre of origin of banana is South East Asia. The banana was grown in Southern Asia even before the prehistoric periods. Historical evidences show that the Arabs have introduced the banana from India to Palestine and Egypt perhaps in seventh century AD. It later spread to East Coast of Africa and subsequently spread throughout Africa. Banana was introduced to Central America in 1516 AD. The spread of banana to West Indian Islands was through Christian Missionaries.

Most of the present-day bananas are thought to have originated from four species, namely, *Musa acuminata* (A), *M. balbisiana (B), M. textiles* (T) and *M. schizocarpa* (S) but with major contributions from the A and B genome species. *M. acuminata* has greater diversity and found in South East Asian countries, Papua New Guinea. It has been classified into subspecies. *M. balbisiana* has less diversity and found in India, Thailand, Philippines, Myanmar and neighbouring countries. It is not divided into subspecies.

4.3 Taxonomy and Classification

The taxonomic position of banana as per phylogenetic system of classification of Hutchinson is:

Phylum	:	Angiospermae
Sub-phylum	:	Monocotyledones
Division	:	Calyciferae
Order	:	Zingiberales
Family	:	Musaceae
Genus	:	*Musa*
Section	:	Eumusa
Species	:	× *paradisiaca*
Binomial name	:	*Musa* × *paradisiaca*

Banana belongs to family Musaceae. Under this family there are two genera, namely, *Musa* and *Ensete.* The cultivated banana (*Musa* × *paradisiaca*) evolved as a result of crossing between two diploid (2n= 2X= 22) wild species, *viz.*, *M. acuminata* (having A genome) and *M. balbisiana* (having B genome) as shown below:

M. acuminata × *M. balbisiana*

↓

Cultivated banana

The *Musa* genus is subdivided into four sections, namely, Eumusa, Rhodochlamys, Australimusa and Callimusa. The genome of Eumusa and Rhodochlamys is of 11 chromosomes and that of Australimusa and Callimusa is of 10 chromosomes. The edible bananas in the section Eumusa had their origin in only two wild species *Musa acuminata* (contributor of 'A' genome) and *Musa balbisiana* (contributor of 'B' genome). Simmonds and Shepherd (1955) used a scoring method to indicate contribution of the two wild species by using 15 characters and the cultivars are classified as AA, AAA, AB, AAB, ABB, AAAA, AAAB, AABB and ABBB genomic groups based on 1-5 numerical scale. Edible bananas have 22, 33 or 44 chromosomes and so the cultivars are diploid, triploid or tetraploid respectively.

4.4 Morphology

4.4.1 Habit: Perennial, herbaceous, succulent plants.

4.4.2 Roots: Adventitious root system arising from the underground rhizome.

4.4.3 Stem: Underground stem called rhizome; leaf sheaths tightly packed together to form pseudostem.

4.4.4 Leaf: Alternate, large, 1-2 m long and 50-70 cm wide; leaf sheath elliptical, packed tightly to form pseudostem; mid-rib concave on the upper surface; veins arising from the midrib long and straight with numerous veinlets. In general leaf is uniformly green with various shades except in few cases where purplish pigmentation is observed on young leaves at sucker stage.

4.4.5 Inflorescence: The plant puts forth an inflorescence after producing 30-60 leaves depending on the cultivar. After the production of all leaves, the plant pushes its inflorescence from the heart. Inflorescence of banana is a complex spike borne at the end of a stout peduncle.

4.4.6 Flower: The flowers are arranged in nodal clusters in two rows, subtended by spathe like bracts, lanceolate to ovate in shape and pinkish green to deep purple black in colour. Banana flowers are arranged either uni- or bi-serially on the nodal cluster on a nodal cushion. They are covered by spathe like bract. Number of flowers varies from1-26 depending on the cultivar. The flowers are of three types; female, bisexual and male. Female flowers appear first, then bisexual and finally male appears.

4.4.7 Perianth: 6 (3+3) or (5) +1, outer three and inner two tepals fused to form the five gamophyllous condition, valvate.

4.4.8 Androecium: 6 stamens, five are fertile; anther bicelled, filaments separate, adnate.

4.4.9 Gynoecium: Ovary inferior, tricarpellary, syncarpous, bilocular, many ovules in axile placentation, style simple and stigma capitate.

4.4.10 Fruit: Berry, fleshy, parthenocarpically developed.

4.4.11 Seed: Absent, if present very small and black.

4.5 Pollination

The edible banana is highly parthenocarpic, where fruits develop without pollination. But, in seeded species/cultivars, pollination is mainly done by bees and birds that visit the flowers for nectar secreted at the tips of ovary and collected at the base of the perianth. The female and male flowers open by 6.30 – 8.00 AM.

4.6 Crossing Techniques

Pollinations are generally carried out between 7.00 and 11.00 AM. Undehisced anthers from male flowers are collected and twisted gently to force them dehisce. Using a soft sable hair brush, the pollen grains are taken out and smeared gently over the stigmatic surface of the female flowers that opened on the day of pollination. The pollinated flowers are to be covered with soft cloth bag.

4.7 Breeding Objectives

- To develop dwarf stature banana suitable for high density planting and to prevent damage through high wind velocity.
- Quality fruits.
- Resistance to biotic and abiotic stresses.
- Wider adaptability.
- Development of male fertile parthenocarpic diploids.
- Longer finger size.
- Suitable for export.
- Keeping quality.

4.8 Breeding Methods

- Introduction
- Hybridization
- Mutation Breeding

4.8.1 Introduction

Introduction of some cultivars was made for resistance to biotic stresses, e.g., Lady Finger from Australia for resistance to bunchy top virus. Cultivar Nain MS from France and Velery from West Indies were introduced for utilization in breeding programme.

4.8.2 Hybridization

In India, breeding work started at Central Banana Research Station, Aduthurai, Tamil Nadu, in 1949. Afterwards, breeding programme was initiated at TNAU, Coimbatore and Kerala Agricultural University, Trichur.

4.8.2.1 Interdiploid Hybridization

Diploids are the main source of genetic variability in breeding new commercial hybrids. Since 1971, interdiploid hybridization was done to synthesize new diploid forms at TNAU, Coimbatore. The breeding objective was to incorporate wilt resistance in Gros Michel using wild *Musa acuminata* diploid as male parent. The tetraploids (progeny of hybrid) were very tall but it was resistant to wilt. The dwarfness was incorporated by using a dwarf mutant High Gate as a substitute for tall Gross Michel. Later work utilized Low Gate which was smaller than High Gate and gave progeny of the same height as Grand Naine.

Matti (AA) cultivar exhibits strong resistance to Sigatoka leaf spot but susceptibility to nematode. Matti is male sterile but sets seeds after pollination and so it can be used as female parent in diploid breeding. The wild species *M. acuminata* ssp. *burmannica* and *M. acuminata* ssp. *malaccensis* have shown resistance to both races 1 and 2 of the Panama wilt fungus, Sigatoka leaf spot and nematode. Other diploid clones involved in diploid male parent synthesis are Anaikomban (AA), resistant to nematode and *Fusarium* wilt but susceptible to Sigatoka leaf spot; Namarai, susceptible to Sigatoka and nematode but tolerant to Panama wilt; Pisanglilin (AA) and Tanget (AA), resistant to Panama wilt and nematode.

4.8.2.2 Triploid Breeding

The crossing of diploid and tetraploid results in the production of a triploid. Tetraploids produce higher proportion of abortive gamete than diploids. So, they may be used as male parent. Production of pollen grains is so great that numerous fertile ones will still be present in spite of the very large majority being non-functional. If tetraploids are used as females, the presence of sterile female gametes would result in drastic reduction in number of seeds. Natural AAA triploid arose from the AA cultivars by chromosomes restitution at meiosis. Several synthetic AAA triploids have been bred using High Gate/ Valery as female parents and wild/edible improved *M. acuminata* diploids as male. For developing AAB triploids, AABB tetraploids are crossed with AA diploids.

4.8.2.3 Tetraploid Breeding

Tetraploids are bred by crossing a triploid female with the improved diploid male. Seed fertile tetraploids are analyzed for potential usefulness in (4X) × (2X) crosses for synthesizing triploid hybrids. Important tetraploids for commercial adoption are FHIA-01 (AAAB), FHIA-02 (AAAA), FHIA-03 (AABB) etc.

4.8.3. Mutation Breeding

Several natural sports of well-established commercial clones have been recognized. High Gate (AAA) is a semi-dwarf mutant of Gros Michel (AAA). Motta Poovan (AAB) is a sport of Poovan (AAB). Ayiranka Rasthali is a sport of Rasthali (or Silk), Barhari Malbhog is a sport of Malbhog, Krishna Vazhai is a natural mutant of Virupakshi and Sombrani Monthan (ABB) is a mutant of Monthan (ABB). Novaria has been derived in Malaysia from gamma-rays irradiation of the clone Grand Naine.

4.8.4. Biotechnological Approaches

Plant tissue culture and molecular biology techniques are applied to enhance the handling and improvement of banana. Important applications are micro-propagation for rapid multiplication and germplasm exchange, embryo culture/rescue for *in vitro* seed germination, cryo-preservation of germplasm and genome manipulation through genetic engineering using cell suspensions or protoplast culture. Somaclonal variation has been successfully applied in Taiwan for the development of improved Cavendish banana cultivars with resistance to *Fusarium* wilt and acceptable fruit quality. The molecular markers are providing tools for phylogenetic investigations and cultivar identification, basic genetic research, marker-assisted selection and diagnostics in pathogen identification.

4.9 Important Banana Clones of India

Important clones of banana in India are given in Table 4.1.

Table 4.1: Important clones of banana in India

Types of banana cultivar	Cultivars	Genome designation	Synonyms	Remarks
Dessert banana	Dwarf Cavendish	AAA	Basrai, Loton (Maharashtra), Kabuli (WB), Bhusaval (AP), Vamanakeli (AP), Pacha Vazhai, Mauritius, Moris (TN)	Second important banana in international trade
	Robusta	AAA	Bombay Green, Harichhal- a semi dwarf mutant of Robusta (Bombay), Robusta (TN), Peddepach Arati (AP)	
	Poovan	AAB	Champa (WB), Karpura Chakkarakeli (AP), Lal Velch (Maharashtra), Palayankodan (Kerala)	Immune to Panama wilt
	Rasthali	AAB	Rasthali (TN), Mutheli (Maharashtra), Malbhog (Bihar), Poovan (Kerala), Amrithapani (AP)	Highly priced banana
	Virupakshi	AAB (Hill banana)	Vella vazhai	
	Chakkarakeli	AAA	Raja Bale (Mysore), Chakkara Kadali (Kerala)	Cultivated in Deccan Plateau
Cooking banana	Monthan	ABB	Monthan (TN), Kanch kola (WB), Khasdi, Bankel (Bombay), Bainsa (Bihar), Batisa (Odisha), Monthan (Kerala)	Leading commercial culinary banana of India.
	Ney Vannan	ABB		
	Nendran	AAB	Ethakai (Kerala), Rajeli (Maharashtra), Kochi kehel (Sri Lanka)	Good chip making quality
Source: Roy 2019				

Review Questions

Part A

Answer the following questions true or false.

1. Monthan is a dessert banana. True/False
2. *Musa* × *paradisiaca* evolved as a result of crossing between *M. acuminata* and *M. balbisiana*. True/False
3. The crossing of diploid and tetraploid results in the production of a hexaploid. True/False
4. The edible banana is highly parthenocarpic. True/False
5. Matti (AA) cultivar shows strong resistance to nematode. True/False

Part B

Answer the following questions.

1. Nendran cultivar has/has not good chips making quality.
2. What is the position of Dwarf Cavendish cultivar in international trade?
3. Why tetraploids are used as male parent?
4. When the breeding work in banana started in India?
5. Robusta is — type of cultivar.

Part C

Write a brief note on each of the following.

1. Triploid breeding in banana.
2. Origin and distribution of banana.
3. Breeding goals in banana.
4. Interdiploid hybridization in banana.
5. Biotechnological approaches in banana.

5

Breeding of Citrus

5.1 Introduction

Citrus is one of the most important fruit crops in the world. It is grown in the tropical and sub-tropical regions of the world. The top ten citrus producing countries are China, India, Nigeria, Brazil, Mexico, USA, Spain, Egypt, Italy and Argentina. The citrus industry in India is the third fruit industry after mango and banana. The important groups of crops in citrus are sweet orange, mandarins, limes, lemons, grape fruits and pummelo. The most important commercial *Citrus* species grown in India are mandarin (*Citrus reticulata* Blanco.), followed by sweet orange (*Citrus sinensis* (L.) Osbeck) and acid lime (*Citrus aurantifolia* Swingle). The major citrus producing states are Maharashtra, Karnataka, Andhra Pradesh and Punjab.

5.2 Origin and Distribution

The information on the origin of different *Citrus* spp. is scarce and only lime is one of the important commercial *Citrus* found growing in wild state. The centres of origin of different *Citrus* spp. are given in the Table 5.1.

Table 5.1: Centre of origin of citrus

Citrus spp.	Centre of Origin
Sweet orange (*C. sinensis* (L.) Osbeck)	Southern China or Indo-China border
Sour orange (*C. aurantium* L.)	South East Asia
Mandarin (*C. reticulata* Blanco.)	Philippines or Indo-China border
Lime (*C. aurantifolia* Christm.)	Northern India and Malaysian archipelago
Pummelo (*C. grandis* (L.) Osbeck)	South-East Asia
Lemon (*C. limon* (L.) Burm.)	South-East Asia
Citron (*C. medica* L.)	South-West Asia and possibly in India

5.3. Taxonomy and Classification

The taxonomic position of *Citrus* spp. as per Hutchinson is:

Phylum	:	Angiospermae
Sub-phylum	:	Dicotyledones
Division	:	Lignosae
Order	:	Rutales
Family	:	Rutaceae
Sub-family	:	Aurantioideae
Tribe	:	Citreae
Sub-tribe	:	Citrinae
Genus	:	*Citrus*
Species	:	number of species
Binomial name	:	*Citrus* spp.

There are six genera under Citrinae, namely, *Citrus, Poncirus, Fortunella, Eremocitrus, Microcitrus* and *Clymenia*. The genera *Poncirus* and *Fortunella* are cold tolerant. Intergeneric and interspecific hybridization has been wide spread in *Citrus*. The GP_1 consists of *C. medica, C. reticulata* and *C. maxima.* The GP_2 consists of other species in the genus. GP_3 consists of other related genera such as *Fortunella, Poncirus, Microcitrus, Eremocitrus* and *Clymenia.*

From various studies, it has been concluded that there are only three distinct taxa:

1. *C. medica*
2. *C. grandis*
3. *C. reticulata*

The genus *Citrus* includes crops such as sweet orange (*C. sinensis*), mandarin (*C. reticulata*), tangerine (*C. reticulata*), grapefruit (*C. paradisi*), lemon (*C. limon*), lime (*C. aurantifolia)*, sour orange (*C. aurantium*) and citron (*C. medica*). The two non-cultivated Indian and Japanese species are *C. indica* and *C. tachibana* respectively. Most species are diploid with 2n= 2x=18. *C. limon* exists in two forms: acid and sweet. Lemon (*C. limon*) is the leading acid citrus fruit (sour) of the world.

5.4 Morphology

A brief description of morphology of three genera — *Citrus, Poncirus* and *Fortunella* is presented below:

5.4.1 Habit: Small, evergreen shrubs or trees up to 10 m or more in height; but *Poncirus* is deciduous.

5.4.2 Root: Usually have a single tap root and lateral roots growing horizontally.

5.4.3 Stem: Single trunk with main branches arising about 1 m above the ground, thorny, thorniness more prominent in young shoots.

5.4.4 Leaf: Compound, trifoliate in *Poncirus*, unifoliate in other genera with articulation between leaflet and petiole, petiole winged, stomata confined to lower leaf surface, oil glands present in palisade layer.

5.4.5 Inflorescence: Solitary or in small clusters of cymes, axillary.

5.4.6 Flower: Bracteate, pedicellate, bisexual, complete, actinomorphic, hypogynous.

5.4.7 Calyx: 5 sepals, gamosepalous, persistent, valvate.

5.4.8 Corolla: 4-8 petals, mostly 5, usually white, fragrant.

5.4.9 Androecium: Stamens 20-40 in groups, anthers 4-celled.

5.4.10 Gynoecium: Superior ovary, 8-15 carpels, syncarpous, each locule containing several ovules in axile placentation, short deciduous style, capitate stigma.

5.4.11 *Fruit*: Hesperidium- superior many celled fleshy fruits.

5.4.12 *Seed*: Few to many, either monoembryonic or polyembryonic.

5.5 Floral Biology

Flower opening starts from morning and extends up to evening, maximum anthesis being between 11.00 am and 12.00 noon. Pollen viability varies from 45-80% depending upon the season. Anther dehiscence takes place either 45 minutes before anthesis or within 45 minutes after anthesis. It varies up to 5 hours after anthesis. The stigma receptivity starts either 15 minutes to 2 hours before anthesis or within 35 minutes to 5 hours after anthesis depending upon the weather. The receptivity lasts for 4-8 days after anthesis.

5.6 Pollination

Citrus is predominantly self-pollinated but some parthenocarpic types are also present. Polyembryony is common in citrus. Partial to complete pollen and ovule sterility and gametophytic self- and cross-compatibility are found in citrus.

5.7 Emasculation and Crossing Techniques

The mature floral buds are emasculated early in the morning and are bagged. The flowers to be used as male parent are bagged the previous evening. The next morning when the anthers dehisce releasing pollens, the male flowers are plucked to pollinate the receptive stigmas of emasculated flowers. The pollinated flowers are bagged and opened after about a week. They are allowed to mature into ripe fruits. Seeds from mature fruits are extracted and sown immediately in sterilized sand and soil mixture. When seedlings are about 15 cm high, hybrid seedlings are identified and others are rejected. The hybrid seedlings are grown to mature trees in the field and the seedlings raised from the fruits are evaluated for suitability as scion or rootstock.

5.8 Breeding Objectives

- To breed cultivars suitable for different agro-ecological situations.
- To breed rootstock having resistance to biotic and abiotic stresses.
- To breed virus free nucellar strains of existing cultivars.
- To breed insect pest and disease resistant cultivars.
- To breed cultivars having less tendency of granulation.
- To breed cultivars suitable for export.

5.9 Breeding Methods

5.9.1 Introduction

It is usually attempted when existing germplasm is felt unsuitable or lack some desirable characters.

- **Mandarin**: Santra from Aurangabad
- **Mandarin**: Kinnow, Cleoptera, Clementine from USA
- **Sweet orange**: Washington Navel, Valencia Late, Jaffa, Blood Red Malta, Tangerines

- **Grape fruit**: Marsh Seedless, Duncan, Foster, Thompson from USA
- **Lemon**: Lisbon, Eureka, Villafranca from USA

Kinnow mandarin was introduced in South India in 1958 and in Punjab during 1959 and has performed extremely well in Punjab and also adopted well in South India.

5.9.2 Clonal Selection

Clonal selection has been very effective within cultivar group. This method involves the survey of different areas, collection of germplasm, raising of germplasm at a place, studying their performance for considerable time and finally selecting superior types. Some examples are:

Orange: Yuvraj blood red, Pusa Sharad, Pusa Round

Mandarin: Madhukhed Seedless

Acid lime: Pramalini, Vikram, Jaydevi (PKM-1), Chakradhar, Tenali, Saisarbati, Pusa Udit, Pusa Abhinav.

Lemon: PAU Selection, Dabwali Baramasi, Pant Lemon, Kagzi Kalan

5.9.3 Hybridization

In Hybridization, the long juvenile phase delays the assessment of the hybrids. It creates problems. In *Citrus*, most cultivars are highly polyembryonic. When we use them as females in crosses, very few weak hybrids are produced. It is difficult to differentiate between hybrid and nucellar seedlings. Electrophoretic techniques separating the isozymes of parents and hybrids may be useful in scion breeding programme as no morphological markers are available at present.

Rootstock hybrids should have desirable traits such as high percentage of nucellar embryony, resistance to different pests, diseases and nematodes. The selected hybrids are evaluated with different scion cultivars and compared with the commercial rootstocks.

In case of scion hybrids, the zygotic seedlings are raised on suitable rootstock. Observations on various traits are recorded. Selected hybrids are evaluated on different rootstocks of different locations and compared with the commercial cultivars.

5.9.3.1 Intergeneric Hybrids

Although intergeneric hybrids are rare in fruit plants, much success have been obtained in *Citrus.*

(a) Hybrids of Poncirus

Citrange: Having the parentage of trifoliate orange (*Poncirus trifoliata*) and sweet orange (*C. sinensis*).

Citrangequat: Trigeneric hybrid of the genera, *viz.*, *Poncirus, Citrus* and *Fortunella.*

Citrangor: By backcrossing Citrange with *C. sinensis.*

Cicitrange: Hybrid between Citrange and *Poncirus trifoliata* × *C. paradisi.*

Citrandarin: Hybrid between *Poncirus trifoliata* and *C. reticulata.*

Citremon: Hybrid between *Poncirus trifoliata* and *C. limon.*

Citradia: Hybrid between *Poncirus trifoliata* and *C. aurantium.*

Citumquat: Hybrid between *Poncirus trifoliata* and *Fortunella japonica* or *F. margarita*

(b) Hybrids of *Fortunella:*

Procimequat: *(F. japonica* × *C. aurantifolia* cv. Mexican*)* × *F. hindisii.*

Limequat: *C. aurantifolia* × *F. japonica.*

Orangequat: *C. reticulata* cv. Satsuma × *F. japonica* × *F. margarita* cv. Meiwa

5.9.3.2 Intrageneric Hybrids

Tangor : *C. sinensis* × *C. reticulata*

Tangelo : *C. reticulata* × *C. paradisi*

Lemonime : *C. limon* × *C. aurantifolia*

Lemonnage : *C. limon* × *C. sinensis*

Lemandarin : *C. limon* × *C. reticulata*

In India, very little work has been done on citrus improvement through hybridization. PKV, Akola has evolved hybrids of Kagzi lime, namely, Hybrid 2, Hybrid 4 and N 52 which are resistant to canker. With regard to rootstock breeding, using trifoliate orange as donor CRH 3, CRH 5 and CRH 41 have been evolved which are resistant to citrus nematode.

An interspecific hybrid between Nepal Round lemon and acid lime has been developed. It is named 'Rasraj' which is resistant to bacterial canker.

5.9.4 Mutation Breeding

Mutation breeding is another important technique to induce mutations for disease resistance, seedlessness etc. by physical and chemical mutagens. Bud variation is an abnormal variation in stem, leaves and fruits that are capable of perpetuating through bud propagation. Such variations in the somatic tissues of the plant are called bud sports or bud mutation. Many of the fruit variations in the *Citrus* have been the result of individual fruit variation in otherwise normal trees. Some examples are given below:

- **Navel orange** : Washington Navel late, Naveline
- **Mandarin** : Clausellina
- **Grapefruit** : Foster, Hudson, Star Ruby, Red Blush, Flame

5.9.5 Polyploidy Breeding

Most of the species and cultivars of *Citrus* are diploid. But polyploidy has been reported in many cultivars. The Hong Kong wild Kumquat (*Fortunella hindisii*) may be the first report of tetraploid. Polyploidy breeding offers the prospect of getting large sized fruit with dwarf plant types. Production of triploids by crossing tetraploid with diploid may be useful in developing seedless cultivars. The seedless lime (*C. latifolia*) is a suitable example.

5.9.6 Biotechnological Approaches

In polyembryonic cultivars the vigorous growth of nucellar embryos inhibits the growth of the zygotic embryo and causes its degeneration prior to seed maturation. Such abortive embryos can be rescued by tissue culture. Tissue culture has effectively been used in obtaining hybrid *Poncirus* plantlet from polyembryonic citrus cultivars. Intergeneric hybridization assisted with tissue culture offers possibility of incorporation of multiple desirable traits from different genera for improvement of citrus rootstocks and scion cultivars. Cell and tissue culture and especially protoplast manipulations have been explored in citrus improvement.

Review Questions

Part A

Answer the following questions true or false.

1. In India, very little work has been done on citrus improvement through hybridization. True/False
2. The information on the origin of different *Citrus* spp. is adequate. True/ False
3. In polyembryonic cultivars the vigorous growth of nucellar embryos inhibits the growth of the zygotic embryo. True/False
4. Most of the species and cultivars of *Citrus* are triploid. True/False
5. In case of scion hybrids, the zygotic seedlings are raised on suitable rootstock. True/False

Part B

Answer the following questions.

1. Give two examples of hybrids of Kagzi lime.
2. The interspecific hybrid 'Rasraj' is resistant to —.
3. What is lemandarin?
4. There are 6/8/9 genera under Citrinae.
5. Orange belongs to family —.

Part C

Write a brief note on each of the following.

1. Emasculation and crossing techniques in *Citrus*.
2. Hybridization in *Citrus*.
3. Mutation breeding in *Citrus*.
4. Breeding goals in *Citrus*.
5. Polyploidy breeding in Citrus.

6

Breeding of Grapes

6.1 Introduction

Grapes are being grown commercially in the tropics. The region between the tropics of Cancer and Capricorn which was once considered to be unsuitable for growing quality grapes is producing good quality grapes. Tropical countries where grapes are grown are India, Indonesia, the Philippines, Thailand, Taiwan, South China, Australia, Brazil, Columbia, Ecuador, Mexico, Venezuela, West Indies, Kenya, Nigeria and Zimbabwe. In India, about 90% of grapes are being produced in the tropical states including Maharashtra, Andhra Pradesh, Karnataka and Tamil Nadu. There are three main commercial uses of grape, *viz*., table, raisin and wine.

6.2 Origin and Distribution

European grape (*Vitis vinifera* L.) is considered to have originated primarily between Caspian and Black Sea region. American grapes belonging to a large number of *Euvitis* and *Muscadinia* species have originated in North America. Wild grapes were present in many places of Eurasia during Neolithic period. It was later spread from Transcaucasia, western Mediterranean countries, Asia Minor, Nile Delta and Balkan Peninsula. Islamic countries promoted for fresh consumption of grapes and its cultivation in Middle East and North Africa. This led to the introduction of table grapes into Spain, France, Italy and New World. In 1525, grapes were planted in Mexico followed by Peru, Chile and Argentina in 1550. Viticulture started in South Africa (1616), California (1697), Australia (1788) and New Zealand (early 19th century).

6.3 Taxonomy and Classification

The taxonomic position of grapes as per Hutchinson is:

Phylum : Angiospermae

Sub-phylum : Dicotyledones

Division : Lignosae

Order : Rhamnales

Family : Vitaceae

Genus : *Vitis*

Species : *vinifera*

Binomial name : *Vitis vinifera* L.

The genus *Vitis* is divided into two subgenera, *viz*., *Euvitis* and *Muscadinia.* There are nearly 57 species under *Euvitis* and three species, *viz*., *rotundifolia, munsoniana* and *popenoel* under *Muscadinia*. The *Euvitis* has a 2n=38 and the *Muscadinia* has 2n=40 chromosomes. The most popular species, *Vitis vinifera* has 2n=2x=40. Among all the species, *Vitis vinifera*, the old-world grape, of the subgenus *Euvitis* is the most important as more than 90% of the grape cultivars grown in the world today are derived directly or indirectly from this species.

6.4 Morphology

6.4.1 Habit: Woody, deciduous vines, climbing by tendril developed from terminal buds.

6.4.2 Root: The grape has spreading (0.60-1.52 m) and descending (1.82-3.65 m) types of root system. The feeder roots are present up to 25 cm depth and up to 60-120 cm away from the trunk.

6.4.3 Stem: Trunk is a permanent stem of the vine on which the whole framework is based. The succulent current season growth arising from a bud after pruning is known as shoot. The woody, mature and dormant shoot is called cane.

***6.4.4 Leaf*:** Smooth shiny leaves with 3, 5 or 7 lobes.

6.4.5 Inflorescence: Axillary and branched raceme (panicle) on current season growth.

***6.4.6 Flower*:** Flowers are small, green, sweetly scented, and borne in clusters. They are perfect, pistillate or staminate.

Perfect flowers: Pistil is functional; stamens are erect and produce fertile pollens.

***Pistillate flowers*:** Pistil is well developed; stamens are reflexed and many produces abundant pollens but remain sterile due to absence of germ pores.

***Staminate flowers*:** Stamens are erect and anthers produce well developed fertile pollen but pistil is only rudimentary without stigma and style with only a small ovary containing incompletely developed ovules.

***6.4.7 Calyx*:** Sepals 5, polysepalous, valvate.

***6.4.8 Corolla*:** Petals 5, polypetalous, calyptras valvate.

***6.4.9 Androecium*:** Stamens 2-7(usually 5), free, anther bilobed, dorsifixed.

***6.4.10 Gynoecium*:** Ovary superior, bicarpellary, syncarpous, bilocular, style simple, stigma bifid.

***6.4.11 Fruit*:** Berry, oval or round with edible skin adhering to flesh.

***6.4.12 Seed*:** Endospermic, seed coat hard.

6.5 Floral Biology

Anthesis occurs most frequently between 6 to 9 AM and may also occur between 2 to 4 PM. Anther dehiscence coincides with anthesis of flowers. In certain cultivars like Angur Kalan and Bhokri, it is observed to be completed well before anthesis. In Gulabi cultivar, dehiscence starts only after anthesis. Stigmatic receptivity is characterized by the presence of sugary secretion on the stigma, giving it a bright appearance. Once stigmatic surface dries, it becomes black, indicating the loss of receptivity. Stigma becomes receptive one day prior to anthesis and remains so, a day after, with maximum receptivity on the day of anthesis.

6.6 Pollination

Vitis spp. had been originally dioecious but later on transformed to hermaphrodite. At present most cultivars of *Vitis vinifera* are hermaphrodite in nature, so self-pollination is a rule.

6.7 Emasculation and Crossing Techniques

The petals are united at the tip. So, the flower does not open from the tip instead the calyptra (corolla) become it detached at the base and develops as a little cap at the time of flowering. For emasculation, the calyptras are carefully lifted with a pair of forceps. It exposes the stamens, stigma etc. Then, the stamens are gently removed before shedding of pollens, several days before the flowers begin to open. After emasculation, the remaining flowers in the bunch are plucked off. Then pollens from the male plant are dusted over the stigma of the emasculated flowers. Pollens may be collected in a vial or the entire cluster (of pollen source) can be cut off for dusting. The pollinated flowers in the cluster are enclosed in a paper bag and tightly secured. Grape seeds are normally extracted manually from ripe berries and are stratified at 4°C for 75 to 90 days before sowing to break seed dormancy. Sand is proved to be a better medium than moss for keeping seeds for stratification. Juvenile period in grape is 3-6 years. It can be shortened by increasing the growth rate of seedlings by growth regulators and use of artificial nutrient medium forcing continuous growth by growing in glass house and top working.

6.8 Breeding Objectives

- Development of high yielding cultivars with desirable quality.
- Breeding for earliness.
- Resistance to downy mildew, powdery mildew, anthracnose and mealy bug.
- Breeding rootstocks resistant to salt and nematodes.
- Breeding cultivars with lesser apical dominance, vigour and capacity to bear on basal nodes.

6.9 Breeding Methods

6.9.1 Introduction

Before 13th century, not much is known about cultivars. Many introductions after 13th century were made and found promising, e.g., Anab-e-Shahi from Middle East in 1890 revolutionized viticulture in India. The cultivars Bhokri and Pandhari Sahabi were introduced 3 centuries ago and they were leading cultivars until late 1950's in Maharashtra.

6.9.2 Selection

6.9.2.1 Selection from Open-Pollinated Seeds

Cheema Sahabi and Selection-94 were developed in Punjab through this method.

6.9.2.2 Clonal Selection

Pusa Seedless (from Thompson Seedless), HS 37-6 (from Perlette), Manjari Naveen (from Centennial Seedless), Tas-e-Ganesh (from Thompson Seedless), Rao Sahebi (from Cheema Sahebi), Sonaka and Manik Champa (both from Thompson Seedless), and Dilkhus (from Anab-e-Shahi) are developed through clonal selection.

6.9.3 Hybridization

In India, hybridization work started in 1958 at IARI, New Delhi. The objective of hybridization was early maturity, high yield, better quality, seedlessness, and resistance to biotic stresses. IIHR, Bangalore started breeding programme in 1968. Objective was to develop superior cultivars for table purpose, raisins, wine and juice.

Pusa Navrang (Madeleine Angevine × Rubired), Pusa Urvashi (Hur × Beauty Seedless), Pusa Aditi (Banqui Abyad × Perlette), Pusa Trishar {(Hur × Bharat Early) × Beauty Seedless} and Pusa Swarnika (Hur × Cardinal) are hybrids released from IARI, New Delhi. The promising hybrids developed at IIHR, Bengaluru are Arka Vati (Black Champa × Thompson Seedless), Arka Kanchan (Anab-e-Shahi × Queen of Vineyards), Arka Shyam (Bangalore Blue × Black Champa), Arka Hans (Bangalore Blue × Anab-e-Shahi), Arka Chitra (Angur Kalan × Anab-e-Shahi), Arka Krishna (Black Champa × Thompson Seedless), Arka Majestic (Angur Kalan × Black Champa), Arka Neelmani (Black Champa × Thompson Seedless), Arka Soma (Anab-e-Shahi × Queen of Vineyards), Arka Trishna (Bangalore Blue × Convent Large Black), Arka Shweta Syn. Shweta Seedless (Anab-e-Shahi × Thompson Seedless) and Arka Neelkiran (Black Champa × Thompson Seedless). National Research Centre on Grapes, Pune has developed a hybrid Medika (Pusa Navrang × Flame Seedless).

6.9.4 Mutation Breeding

Mutation breeding may be attempted as a complementary tool in grape breeding for one or more important traits, without changing the whole genetical set up. Induced mutations have resulted in a few improved cultivars such as New Perlette (Loose Perlette) from Perlette, Red Niagara from Niagara, Robin Cardinal from Cardinal.

6.9.5 Polyploidy Breeding

Polyploidy breeding has great importance in the improvement of table grapes. Main benefit from polyploidy is the increase in fruit size. Autotetraploids are considerably sterile and less productive than the parents. New triploid seedless grapes may be developed from crossing of diploids and induced tetraploids. The triploids are highly sterile. Colchicine is generally used as an aqueous solution of 0.25-0.50 % with 5-10 % glycerin to induce polyploidy. But in the development of tetraploids, the unaffected diploid cells may try to suppress the development of tetraploid shoots. The best method is to retain only single top bud on a shoot removing other buds. Marvel Seedless from Delight, Early Giant from Campbell, Muscat Common Hall from Muscat Alexandria, Black

King from Campbell, Wallis Giant from Concord, Case from Sultana etc. are important examples of polyploidy.

6.10 Cultivars Released

Indian Agricultural Research Institute (IARI), New Delhi; Indian Institute of Horticultural Research (IIHR), Bengaluru; National Research Centre on Grapes, Pune and State Agricultural Universities have developed several cultivars/hybrids for table and juice purposes. Salient characteristics of some them are given below:

Hybrids

Arka Vati: Berries are medium, yellowish-green, ellipsoidal-spherical, sweet, seedless with 22-25% TSS. It is multipurpose cultivar used for making raisins and wine.

Arka Hans: Berries are yellowish-green, ellipsoidal-spherical, seeded with pleasant flavour, TSS 18-21%, used for wine making.

Arka Kanchan: Late maturing, berries are golden yellow, ellipsoidal to ovoid, seeded with pleasant muscat flavour, TSS 19-22%, suitable for table purpose and wine making, high yield potential, poor keeping quality.

Arka Shyam: Berries are medium large, shiny black, spherical to ovoid, seeded, with mild flavour, resistant to anthracnose, suitable for wine making and table purpose.

Arka Neelamani: Berries are black, seedless with crispy pulp, TSS 20-22%, tolerant to anthracnose, average yield 28 t/ha, suitable for table purpose and wine making.

Arka Shweta: Berries are yellow, ovoid, seedless, TSS 18-19%, average yield 31 t/ha, used for table purpose, export potential good.

Arka Majestic: Berries are deep tan coloured, uniform round, seeded, TSS 18-20%, tolerant to anthracnose, average yield 38 t/ha, export potential good.

Arka Chitra: Berries are golden-yellow, with pink blush, slightly elongated, TSS 20-21%, tolerant to powdery mildew, average yield 38 t/ha, suitable for table purpose.

Arka Krishna: Berries are dark coloured, seedless, round to ovoid, TSS 20-21%, average yield 33 t/ha, suitable for juice making.

Arka Soma: Berries are greenish-yellow, round to ovoid, pulp meaty with

Muscat flavour, TSS 20-21%, tolerant to anthracnose, downy mildew and powdery mildew; average yield 40 t/ha, good for preparing white dessert wine.

Arka Trishna: Berries are deep tan in colour, round to ovoid, TSS 22-23%, resistant to anthracnose, tolerant to downy mildew, average yield 26 t/ha, good for wine making.

Punjab Purple: It is a hybrid of Catawba × Beauty Seedless. It ripens in first week of June and is rich in anthocyanin —a source of antioxidants and is suitable for processing into juice, nectar and ready for serve beverage. It is black in colour and juicy. Potential yield 24 to 52 t /ha under Y-system, TSS 17.7%.

Pusa Navrang: Early ripening (1st week of June), basal bearer, containing red pigment both in peel and pulp; bunch loose, medium size; berries round and medium sized, TSS 18-20%, suited for coloured juice and wine making; resistant to anthracnose.

Pusa Urvashi: Early ripening (1st week of June), basal bearer, bunch loose and medium sized; berries greenish-yellow and seedless; TSS 20-22%; suitable for table purpose and raisin making.

Pusa Aditi: Early maturing (1st week of June), berries yellowish-green, round, large (2.7g), seedless, pulp firm, TSS 19.3°Brix, tolerant to anthracnose and powdery mildew. Grape bunches and berries are GA_3 responsive. Good for table purpose and making juice; average yield 12-15 t /ha.

Pusa Trishar: Early maturing, suitable for sub-tropical conditions, semi-vigorous and spur-pruned, maturing by 10th June; berries round (2.15g), yellowish-green, TSS 18.4° Brix, pulp firm; grape bunches and berries GA_3 responsive; good for table purpose and juice making; average yield 14-16 t/ha

Pusa Swarnika: Large berry, round, golden yellow, firm pulp, very sweet, TSS 20-22° Brix; natural loose bunches; average bunch size 409 g and bunch length 20 cm; fruits ready for harvest between 75-80 days after full bloom; suitable for table purpose, juice making and Munakka preparation.

Cultivars

Cheema Sahebi: A selection (Sel 7) made by Dr. G.S. Cheema from open pollinated seedlings of Pandhari Sahebi at Ganeshkhind Nursery, Pune. It is a vigorous and a very heavy yielder though very late in ripening.

Kali Sahebi: Berries large, oval-cylindrical, reddish-purple and seeded; TSS 22%; susceptible to rust and downy mildew, average yield 12-18 t /ha; suitable for table purpose.

Manjari Naveen: Released by NRC on Grapes, Pune, TSS 16-18° Brix, total acidity 0.45-0.55%, early maturing cultivar; fruits ready for harvest between 115-120 days after pruning.

Pusa Seedless: Resemble its parent Thompson Seedless with regard to most of the characters, but its berries are more elongated; highly responsive to GA_3 application, TSS 22-24° Brix; suitable for table purpose and raisin making.

Sharad Seedless: Clonal selection from Kishmish Cherney; berries seedless, black, crisp and very sweet, TSS 18-21° Brix, responds well to GA_3, good shelf life, berries oblong to elliptical, bluish black, acidity 0.5-0.7%, grown for table purpose.

Review Questions

Part A

Answer the following questions true or false.

1. The genus *Vitis* is divided into three subgenera. True/False
2. Pusa Navrang is a hybrid. True/False
3. In grapes the petals are free at the tip. True/False
4. Anthesis in grapes occurs most frequently between 10 to 11 am. True/False
5. There are three main commercial uses of grape. True/False

Part B

Answer the following questions.

1. Mention two hybrids of grapes.
2. Where is the centre of origin of *Vitis vinifera*?
3. Pusa Seedless is a clonal selection from —.
4. In India, hybridization work in grapes started in 1958 at —.
5. Anab-e-Shahi was introduced from the Middle/Far East in 1890.

Part C

Write a brief note on each of the following.

1. Emasculation and crossing technique in grapes.
2. Floral biology and pollination in grapes.
3. Hybridization in grapes.
4. Origin and distribution of grapes.
5. Breeding objectives of grapes.

7

Breeding of Papaya

7.1 Introduction

Papaya is a fruit crop grown in kitchen garden, backyards of home and orchards. It is grown as a filler crop in orchard. India is the largest producer of papaya in the world. It is also cultivated in Brazil, Mexico, Australia, Hawaii, Malaysia, Taiwan, Peru, Florida, Gold Coast, South Africa and Bangladesh. In India, it is widely cultivated in Uttar Pradesh, Bihar, Karnataka, Jharkhand and Madhya Pradesh. The ripe fruits of papaya are consumed throughout the tropics and sub-tropics. Fruits are also used in the preparation of jam, soft drinks, ice cream flavour, crystallized fruits and syrups. Unripe fruits are commonly used as vegetables. Papain is prepared from the latex of immature fruits. It is a proteolytic enzyme used for tendering of meat, preparation of chewing gum and cosmetics. The ripe fresh fruit is rich source of vitamin A, carbohydrates and minerals. The fruits are prescribed for piles, dyspepsia, liver, spleen and digestive disorders. Roots are used as nerve tonic. The plants flower five months after planting and fruits are ready five months after flowering. So, it has attracted attention of fruit growers to cultivate on a large scale in the country.

7.2 Origin and Distribution

Papaya (*Carica papaya* L.) is native to tropical America. It is believed to have originated in Southern Mexico (de Candolle, 1884). It was introduced into India in the early part of 16th century from the Philippines through Malaysia. Opinions differ on the origin of *C. papaya* in tropical America. It is likely that papaya originated in the low lands of eastern Central America, from Mexico to Panama. The seeds were distributed to the Caribbean and South East Asia during the 16th century, and to India and Africa.

7.3 Taxonomy and Classification

The taxonomic position of papaya as per Hutchinson system is:

Phylum : Angiospermae

Sub-phylum : Dicotyledones

Division : Lignosae

Order : Cucurbitales

Family : Caricaceae

Genus : *Carica*

Species : *papaya*

Binomial name : *Carica papaya* L.

Carica papaya L. is diploid (2n=2x=18). Earlier 31 species were reported in the genus *Carica,* but recently another taxonomic revision was proposed and supported by molecular evidence that genetic distance was found between papaya and other related species. The Caricaceae family consists of 6 genera and 35 species. The genera belonging to Caricaceae family are *Carica* (1 species), *Jarilla* (3 species), *Horovitzia* (1 species), *Jacaratia* (7 species), *Vasconcellea* (21 species), and *Cylicomorpha* (2 species).

7.4 Morphology

7.4.1 Habit: A small tree, herbaceous, evergreen, fragile, perennial, 1-10 m high.

7.4.2 Root: Tap root system.

***7.4.3 Stem*:** Hollow and soft wooded, usually unbranched while young but at later stage upright shoots develop, marked with scars of fallen leaves.

***7.4.4 Leaf*:** Large, deeply lobed, alternate, palmately lobed, mostly long petiolate, stipulate, each lobe further pinnately lobed, petiole long, hollow, 6 cm or more in length, venation reticulate, multicostate.

***7.4.5 Inflorescence*:** Papaya is dioecious but hermaphrodite forms occur. The inflorescence is axillary in position. Flower are borne in the leaf axil singly or in pendent panicles. Since papaya is a polygamous species, many forms have been reported.

***7.4.6 Flower*:** In general, there are three types of flowers, *viz*., (i) staminate (male) (ii) pistillate (female) and (iii) hermaphrodite (bisexual).

(i) Staminate Flower: Bracteate, bracts leafy or scaly; bracteolate, sessile, in cluster or raceme, incomplete, actinomorphic, funnel shaped, 2.5-3.2 cm long.

Calyx: Sepals 5, gamosepalous, small, light green.

Corolla: Petals 5, gamopetalous, tube like elongated, yellow.

Androecium: Stamens 10, in two whorls, inner ones smaller, epipetalous, anthers bilocular, dorsifixed, introse.

(ii) Pistillate Flower: Bracteate, bracts leafy or scaly; bracteolate, sessile, in cluster or raceme, incomplete, actinomorphic, funnel shaped, 2.5-3.2 cm long.

Calyx: Sepals 5, gamosepalous, light green.

Corolla: Petals 5, linear, polypetalous, twisted aestivation.

Gynoecium: Sessile, superior ovary, pentacarpellary, syncarpous, pentalocular, many ovules in parietal placentation, style very short, stigma 5, dilated or linear, simple or lobed.

Fruit: Fleshy berry, shape ovoid-oblong to nearly spherical, skin smooth, green turning yellowish or orange when ripe; flesh orange or reddish orange, sometimes yellowish; central cavity 5-angled, with numerous seeds

Seeds: Spherical, black or grayish, wrinkled, enclosed in gelatinous sarcotesta formed from the outer integument, straight embryo, cotyledons oblong and flat

(iii) Hermaphrodite Flower: Flowers similar to pistillate type but inflorescence is multi-flowered (5-6 flower corymb), corolla gamopetalous, stamens 10 (5+5), sessile at the base of petals, ovary usually functional.

7.5 Floral Biology

The peak period of anthesis is between 5.00- 6.00 AM. The dehiscence of anther in staminate and hermaphrodite flowers takes place between 10 to 12 hours before anthesis. High temperature and low humidity hasten the time of anther dehiscence. In papaya, high percent pollen germination can be achieved with 5% sucrose solution. Papaya pollen can be stored up to 5 years without losing the viability if proper conditions are created for storing. Stigma receptivity is 48 hours before the anthesis and continues to be receptive up to 72 hours of anthesis and stigma receptivity is maximum on the day of anthesis.

7.6 Pollination

Carica papaya is a cross-pollinated crop. The cross-pollination is done by honey bees. To ensure good fruit set in the dioecious cultivars of papaya (e.g., Pusa Giant, Pusa Dwarf, Pusa Nanha, CO-1, CO-2, CO-5, CO-6 etc.) minimum 10 % male plants should be planted in the orchard. However, the pollinizers are not required in gynodioecious cultivars (e.g., Pusa Majesty, Pusa Delicious, Solo, Sunrise Solo, Coorg Honey Dew etc.).

7.7 Emasculation and Crossing Techniques

For breeding purpose, the terminal buds are selected and tagged; then the selected buds are emasculated before anthesis. Anther dehiscence/pollens shedding begins six hours before anthesis. The pollen grains from hermaphrodite flowers are highly fertile. It has been observed that when the female flowers are pollinated with pollens obtained from bisexual tree, will set the maximum number of fruits. Dioecious papaya produces male and female flowers separately on different plants while gynodioecious cultivar produces both female and hermaphrodite flowers on the same plants. Hence, in case of dioecious papaya, there is no need of emasculation and the female flower can be pollinated by pollen from male parent or stored pollen. In case of gynodioecious papaya, there is need of emasculation.

7.8 Breeding Objectives

- To develop cultivars of dwarf stature, smaller petiole length and early bearing.
- To evolve cultivars with high yield and good quality fruits.
- To develop cultivars with low seed cavity and more pulp thickness.
- To develop cultivars having good keeping quality and suitable for export.
- To breed cultivars for high latex yield for papain production.
- To develop cultivars resistant against biotic and abiotic stresses.

7.9 Breeding Methods

7.9.1 Introduction

Exotic cultivars such as Sunrise Solo, Red Lady and Sinta have been introduced into India. Sunrise Solo is precocious, low bearing and orange red flesh cultivar. Red Lady is vigorous and tolerant to papaya ring spot virus. Sinta is gynodioecious, semi-dwarf and tolerant to PRSV-p. The plant bears higher quality fruits with firm flesh and also produces good quantity of papain.

7.9.2 Selection

Pusa Delicious, Pusa Majesty and Coorg Honey Dew are examples of some cultivars selected with characters like gynodioecious. Other selections like Pusa Giant, Pusa Dwarf, CO-6 and Punjab Sweet are dioecious in nature. Other traits which were kept in mind for selection of superior papaya cultivars are high yield, good quality, high papain production, dwarfness, good keeping quality and resistance to biotic and abiotic stresses.

7.9.3 Hybridization

Inter-varietal and interspecific hybridization have been carried out in papaya to enhance yield, quality, resistance to biotic and abiotic stresses. Many intervarietal hybridization programmes have been initiated in the world and in India for fruit and papain production. CO 3, CO 4, CO 7, HPSC-3 and Arka Surya are the cultivars released as a result of intervarietal hybridization. Most of the interspecific hybridization programmes have focused on the introgression of resistance to PRSV-p. Till date many researchers have carried out interspecific hybridization including different species of *Vasconcellea* to incorporate PRSV-p resistance and cold tolerance.

7.9.4 Mutation Breeding

The work carried out at Pusa, Bihar has resulted in the release of one dwarf mutant "Pusa Nanha". Ram and Srivastava (1984) exposed dry seed samples of Pusa 1-15 to various doses of gamma-rays. They observed that the LD_{50} seems to lie between 20-25 kr. In M_6 generation, uniformity was obtained for plant height, and fruit size and shape. It was later released as Pusa Nanha.

7.9.5 Polyploidy Breeding

Triploid plants are seed sterile. So, in cases where the seedlessness is employed to improve the quality of fruits, such as in banana, apple, citrus, grapes, papaya, etc, induction of triploid plants would be of great use.

Consumers prefer seedless fruit. Seedless triploid papaya fruits over their diploid counterparts are that they also have smaller seed cavity. Thus, they are expected to be more attractive to consumers for direct consumption and processing.

7.10 Cultivars Released

Washington: Table purpose cultivar, fruits round to ovate, medium-large, few seeds, skin bright yellow colour at ripening; average weight of fruit 1.5-2.0 kg; dioecious.

Coorg Honey Dew: Gynodioecious selection from Honey Dew at Coorg.

CO 1: Plant dwarf; fruit round; selected from cultivar Ranchi.

CO 2: Pure line selection from local type, suitable for papain production.

CO 3: Hybrid between CO 2 × Sunrise Solo; plant vigorous; fruit medium sized with good keeping quality.

CO 4: Hybrid between CO 1 × Washington; fruit large; keeping quality good; flesh colour yellow.

CO 5: Selection from Washington; good for papain production.

CO 6: Selection from Pusa Majesty; plant dioecious and dwarf; suitable for papain production.

Pusa Majesty: Selection from cultivar Ranchi; Gynodioecious, suitable for high papain.

Pusa Delicious: Gynodioecious selection from cv Ranchi; good fruit quality.

Pusa Giant: Dioecious selection from cv Ranchi, tolerant to strong wind.

Pusa Dwarf: Selection from cv Ranchi; plant dwarf and dioecious, fruit shape oval.

Pusa Nanha: Developed through mutation breeding, plant dwarf, suitable for high density planting; fruit quality good.

Pant Papaya-1: Dwarf plant type; fruit medium sized (1-1.5 kg).

Pant Papaya-2: Plants vigorous; fruit (1-2 kg) of good quality.

Pant Papaya-3: Vigorous, medium sized fruits (0.5-0.9 kg) of good quality.

Punjab Sweet: Frost tolerant and dioecious in nature.

Surya: Gynodioecious hybrid (Sunrise solo × Pink Flesh Sweet); skin of fruit smooth becoming uniformly yellow on ripening; medium sized fruits (0.6-0.8 kg) with small fruit cavity; pulp deep red and sweet; keeping quality of fruits good.

HPSC-3: Hybrid between Tripura Local × Honey Dew, resistant to papaya mosaic potex virus.

Arka Prabhath: It is developed from cross between (Surya × Tainung-1) × Local Dwarf; gynodioecious; semi-vigorous; pulp firm and deep pink, fruit weight is 0.9-1.2 kg. Keeping quality is good.

Review Questions

Part A

Answer the following questions true or false.

1. *Carica papaya* is a self-pollinated crop. True/False
2. For breeding purpose, the terminal buds are selected. True/False
3. Surya is a dioecious hybrid. True/False
4. Pusa Nanha is developed through mutation breeding. True/False
5. The peak period of anthesis in papaya is between 9.00- 10.00 am. True/ False

Part B

Answer the following questions.

1. Pusa Delicious is developed through —.
2. What is the difference between dioecious and gynodioecious cultivars?
3. *Carica papaya* L. is believed to have originated in Northern/Southern Mexico.
4. Papaya belongs to Order —.
5. Name two gynodioecious cultivars of papaya.

Part C

Write a brief note on each of the following.

1. Floral biology of papaya.
2. Mutation breeding in papaya.
3. Emasculation and crossing techniques in papaya.

8

Breeding of Guava

8.1 Introduction

Guava is the fourth important fruit crop next to mango, banana and citrus. It is known as the 'Apple of tropics'. It is very rich and cheap source of Vitamin C and also contains a fair amount of calcium. Important guava growing states are Uttar Pradesh, Bihar, Madhya Pradesh and Maharashtra. Allahabad district of Uttar Pradesh has the reputation of growing the best quality of guava fruits in the world. The importance of guava is due to the fact that it is the hardy fruit which can be grown in alkaline and poorly drained soil. Guava fruits are used for making jam, jelly and various culinary purposes. The fruit can be canned in sugar syrup.

8.2 Origin and Distribution

Tropical America is supposed to be the centre of origin of guava where it is found in wild as well as cultivated forms. Guava was introduced into India at a very early time before 17th century.

8.3 Taxonomy and Classification

The taxonomic position of guava as per Hutchinson system is:

Phylum : Angiospermae

Sub-phylum : Dicotyledones

Division : Lignosae

Order : Myrtales

Family : Myrtaceae

Genus : *Psidium*

Species : *guajava*

Binomial name: *Psidium guajava* L.

Most of the commercial cultivars are diploids (2n=22), while the seedless ones are triploids. Genus *Psidium* contains about 150 species.

8.4 Morphology

***8.4.1 Habit*:** Evergreen, shrub or small tree, sometimes growing up to 8-9 m.

***8.4.2 Root*:** Tap root system, shallow root system.

***8.4.3 Stem*:** Young shoots angular and ridged, pubescent, bark is smooth, grayish brown and peeling off thin flakes.

***8.4.4 Leaf*:** Petiolate, petiole short, 3-10 mm long; exstipulate, simple; opposite phyllotaxy, blade oblong to elliptic, 5-15 × 4-6 cm; margin entire; apex obtuse to bluntly acuminate; somewhat thick and leathery, dull grey to yellow green above, slightly downy below, finely pubescent below, veins prominent.

***8.4.5 Inflorescence*:** Axillary, solitary or cymes of 2-3 flowers on the current season's growth.

***8.4.6 Flower*:** Pedicellate, pedicel about 2 cm long; bracteate, bracts 2, linear, bisexual, white, epigynous.

***8.4.7 Calyx*:** Entire in bud, splitting irregularly into 4-6 lobes, persistent.

***8.4.8 Corolla*:** 5, polypetalous, imbricate aestivation.

***8.4.9 Androecium*:** Stamens numerous, inserted in rows on disc; filaments white, anthers pale yellow.

***8.4.10 Gynoecium*:** Ovary inferior, ovules numerous, style filiform, greenish yellow, long exerted above stamens; stigma capitate.

***8.4.11 Fruit*:** Berry, surmounted by calyx lobes; shape round, ovoid or pyriform, 2.5-10 cm long; colour pale green to bright yellow or red; flesh white, yellow, pink or red. The flavour is sweet, musky and the ripe fruit is aromatic in high degree.

***8.4.12 Seed*:** Numerous, small, hard, reniform, embedded in soft flesh towards the centre of the fruit.

8.5 Floral Biology

Peak time of anthesis is between 5.00- 6.30 AM in most of the cultivars of guava. The dehiscence of anthers starts at 15-30 minutes after anthesis and continues for two hours. The pollen fertility is high in almost all the cultivars. The pollen fertility is 78% and 91% in Allahabad Round and Lucknow Safeda, respectively. The receptivity of the stigma is found to be maximum during anthesis. Stigma is found to be receptive two days before dehiscence, extending up to 4 days.

8.6 Pollination

Although guava is self-pollinated crop, some extent of cross-pollination takes place by insect.

8.7 Emasculation and Crossing Techniques

In guava, flowers selected for crossing are emasculated when at the 'calyx break stage', a day before opening. Pollen from the male parent is brought from an unopened flower, preferably at calyx break stage. The stigmatic surface is gently smeared with the pollens and flowers are bagged. Under Bengaluru condition, pollination carried out during the morning hours between 10 AM to 12 noon has given better results.

8.8 Breeding Objectives

Scion Improvement

- Dwarf plant stature.
- Tenete branching habit for more fruiting area.
- Development of seedless cultivar.
- Less pectin content for edible purpose.
- More pectin for processing.
- Uniform ripening.
- High keeping quality.
- Better fruit quality in terms of shape, size, colour, firmness, thick pulp and pleasant aroma.
- To develop cultivars with less number of soft seeds.
- Resistance to wilt, tea mosquito bug, fruit fly etc.

Rootstock Improvement

- Wilt resistance.
- Dwarfness.
- Tolerance to various biotic and abiotic stresses.
- Drought tolerance.

8.9 Breeding Methods

8.9.1 Introduction

Several introductions of promising genotypes have been made in guava growing countries. In India, many introductions made from Hawaii, Brazil, Thailand are being cultivated and used in breeding programmes. Similarly, introductions of Indian cultivars like Allahabad Safeda and Sardar have given excellent results in other parts of the world.

8.9.2 Selection

In guava, majority of successful cultivars are released through selection. Important cultivars released through selections are: Lucknow-49 (Sardar), CISH-G-1, CISH-G-2, CISH-G-3 (Lalit), CISH-G-4(Sweta), Arka Mridula (Sel-8), Allahabad Surkha, Pant Prabhat, Dhareedar, Banarasi Surkha, Bangalore Local.

8.9.3 Hybridization

In general, intervarietal crosses in guava are successful, having no crossability barriers. However, varietal cross incompatibility in guava is reported in Behat Coconut × Sardar, and Behat Coconut × Apple Colour. In India, breeding work for guava improvement has been going on at several research institutions and very few guava hybrids were released over the years and important ones are: Safed Jam, Kohir Safeda, Arka Amulya, Arka Kiran, Hisar Safeda, Hisar Surkha.

8.9.4 Polyploidy Breeding

The basic chromosome number of *P. guajava* is X=11. Mainly, seeded types are diploids and the highly seedless types are triploids. Seedless type, or less and soft seeded type is an important commercial attribute. Commercially recognized cultivars as seedless are not completely seedless and types ranging from seedless to less-seeded can be identified. Several parts of our country reported seedless cultivars. Among polyploidy, triploids are the most important as they produce seedless fruits. Natural auto-tetraploid in guava is also reported. Tetraploidy can be induced by colchicine treatment. Chromosome set of Indonesian Seedless and Bangkok Apple is triploid. A promising tetrasomic dwarfing rootstock (Aneuploid No.82, renamed as Pusa Srijan) has been identified. Based on studies on the effect of Pusa Srijan rootstock on Allahabad Safeda, it can be used as dwarfing rootstock on commercial scale, for increasing production and profitability of guava orchards.

8.9.5 Mutation Breeding

Naturally occurring mutations are not rare in guava. The effect of gamma rays (1, 2, 3, 4 and 5 kR) on buds of guava cv Sardar was studied. After the treatment, these were budded onto Lucknow-49 rootstock. Variability for plant height, intermodal length and stem diameter was maximum in 2kR treatment; while, for number of branches, number of leaves and breadth of leaves, maximum variability was noted in 4, 1 and 3 kR treatments, respectively. Mutagenic treatment had no significant effect on stomatal size.

8.10 Cultivars Released

A. Cultivars

Allahabad Safeda: This is the famous cultivar of Uttar Pradesh. Plant is dwarf with compact, globose, round crown. It has profuse branching with dense foliage. Fruits are medium to big in size, skin smooth, flesh white, soft and firm, roundish fruit shape, the average fruit weight is about 185 g, contains approximately 300 to 315 viable seeds. Taste is sweet with excellent fruit quality. It takes about 60-75 days for fruit maturity, keeping quality of fruit is good.

Sardar Guava: (Syn: Sardar, L-49 or Lucnknow-49). This cultivar was developed by Cheema and Deshmukh (1927) through selection from Allahabad Safeda. It is a semi-dwarf tree, 2-3.5 m tall, profuse branching type, somewhat flattened top, dense foliage, large leaf, elliptic ovate to oblong in shape, apex obtuse, base round, fruit medium to large, average fruit weight about 175 g, skin colour primrose with occasional red spot on the skin, taste sweet and keeping quality excellent.

Chittidar: The cultivar is similar to Allahabad Safeda but the fruits of this cultivar are highly red dotted on the surface. Plant is tall with round crown and spreading branches. The cultivar was developed in Allahabad region. Its leaves are large, elliptic-ovate to oblong-elliptic in shape; apex acutely pointed with roundish base. Fruits are almost round, white flesh with smooth skin. Keeping quality of fruit is good.

Sebia (Syn. Appple Colour): This cultivar is famous for its skin colour but commercially not cultivated due to poor yield. Tree is medium in height, with broad crown and spreading growth habit. This cultivar has also originated in Allahabad district of Uttar Pradesh. Fruits are pink in colour with small dots on the surface. Shape is spherical, sweet in taste with good keeping quality.

Red Fleshed (Syn. Red Fleshed Local, Hafsi): Tree is medium tall with spreading branches. The flesh of the fruit is red with good taste and flavour. The number of seeds is less, fruit size is medium and fruit shape is spherical to roundish ovate. The keeping quality is poor.

Seedless: In this group two cultivars are famous, namely, Nagpur Seedless and Saharanpur Seedless. Both the cultivars may be same. Two types of fruits are found, i.e., complete seedless and partial seedless. The trees are tall with upright branches. Fruits are oblong in shape, straw yellow in colour with thick and creamy white flesh.

Baruipur: Tree is medium to tall, with spreading growth habit. Its crown is broad and compact. Fruit is roundish, yellow in colour with white flesh. Keeping quality of fruit is medium. This is the most important commercial cultivar of West Bengal.

Dhawal: Released by CISH, Lucknow; heavy bearer (about 20% higher than Allahabad Safeda); fruits light yellow on ripening, soft seeded and white pulp; taste sweet with muskiness.

Lalima: Released by CISH, Lucknow; attractive crimson colour fruits, higher shelf life.

Lalit: Selection from half-sib population apple guava, suitable for both fresh and processing purposes, released by CISH, Lucknow, saffron yellow coloured fruits with red blush; pulp firm and pink, good blend of sugar and acid.

Shweta: Selection from half-sib population of apple guava cultivar; fruits globose, medium size, weight 225 g; skin creamy white; flesh snow white; TSS 12.5-13.2%; good keeping quality.

Arka Mridula: Selection from open-pollinated seedlings of Allahabad Safeda; plants semi-vigorous and spreading; fruits round, about 180 g, skin yellow and smooth, flesh white, TSS around 12^0 Brix, soft seeded; keeping quality good; pectin content 1.041%, good for jelly making.

B. Hybrids

Safed Jam: Hybrid between Allahabad Safeda × Kohir; developed at FRS, Sengareddy; tree medium tall; heavy yielder, fruits round, large; skin thin, good taste, few soft seeds.

Kohir Safeda: Hybrid between a selected, heavy yielding line of Kohir and Allahabad Safeda; tree vigorous, fairly large and dome shaped; fruits large, few soft seeds; white pulp.

Arka Amulya: Hybrid between Allahabad Safeda and Triploid; developed at IIHR, Bengaluru; plants semi-vigorous and spreading type; fruits medium sized (180-200 g); pulp white; TSS 12.5^0 Brix; good keeping quality.

Arka Kiran: Hybrid between Kamsari and Purple Local; developed at IIHR, Bengaluru; plants semi-vigorous, amenable to high density planting; fruits sub-globose, about 200-230 g; pulp deep pink, thick with good flavour; seeds medium soft; TSS 12.0-12.5^0 Brix.

Arka Rashmi: Hybrid between Kamsari and Purple Local; developed at IIHR, Bengalure; fruits medium sized (180-200 g), globose; pulp deep pink, medium soft seeds, TSS 12^0 Brix; excellent taste.

Hisar Safeda: Hybrid between Allahabad Safeda and Seedless; developed at CCSHAU, Hisar; plants upright, compact crown; fruits round with a smooth surface, skin creamy yellow; fruit average weight 92 g; pulp creamy white; few soft seeds; TSS 13.4%.

Hisar Surkha: Hybrid between Apple Colour and Banarasi Surkha; developed at CCSHAU, Hisar; tree crown is broad to compact; fruits round skin yellow with red dots; average fruit weight 86 g; pulp pink, TSS high (13.6^o Brix).

Review Questions

Part A

Answer the following questions true or false.

1. Tropical America is supposed to be the centre of origin of guava. True/False
2. The basic chromosome number of *P. guajava* is X=12. True/False
3. Arka Rashmi is a hybrid. True/False
4. Naturally occurring mutations are rare in guava. True/False
5. In general, inter-varietal crosses in guava are successful. True/False

Part B

Answer the following questions.

1. Guava is a self-/cross-pollinated crop.
2. Name two cultivars developed by selection.
3. Peak time of anthesis is between — and — AM in most of the cultivars of guava.

4. Write the main characters of Allahabad Safeda cultivar.
5. When was guava introduced into India?

Part C

Write a brief note on each of the following.

1. Emasculation and crossing techniques in guava.
2. Breeding objectives of guava.
3. Floral biology of guava.
4. Polyploidy breeding in guava.
5. Selection and hybridization in guava.

9

Breeding of Pineapple

9.1 Introduction

Pineapple is one of the most important commercial fruits of the world. It is good source of vitamin A, B and fairly rich in vitamin C, minerals like calcium, phosphorous and iron. Fresh fruit is used as dessert. Fruit is also used for juice, squash, jam, mixed jam and in canned form. Fruit core is used for preparing candy. Some cultivars of pineapple are used for extraction of fibre. The important pineapple growing states are Assam, West Bengal, Tripura, Kerala, Andhra Pradesh, Bihar, Karnataka and Madhya Pradesh.

9.2 Origin and Distribution

Pineapple is believed to have originated in Brazil. The wild Brazilian pineapple (*Ananas microstachys* Lindle) is considered as ancestor of cultivated pineapple. It reached India during 1548.

9.3 Taxonomy and Classification

As per Hutchinson the taxonomic position of pineapple is:

Phylum : Angiospermae

Sub-phylum : Monocotyledones

Division : Calyciferae

Order : Bromeliales

Family : Bromeliaceae

Genus : *Ananas*

Species : *comosus*

Binomial name : *Ananas comosus* (L.) Merr.

Ananas comosus is diploid having chromosome number 2n = 2x =50.

9.4 Morphology

***9.4.1 Habit*:** Robust, stoloniferous, gregarious herb,50-125 cm high.

***9.4.2 Root*:** Shallow root system.

***9.4.3 Stem*:** Stem is short and thick, 15-25 cm long, narrow at the base and wider at the top with short internodes.

***9.4.4 Leaf*:** Sessile, numerous, linear, thickly coriaceous; crowded on stem in a bushy rosette; spine present at the tip, smooth or serrate; upper leaf surface dark green or mottled red.

***9.4.5 Inflorescence*:** Apical meristem broadens to form a compact inflorescence.

***9.4.6 Flower*:** Bracteate, 100-200 reddish purple flowers, regular, bisexual, epigynous; flowers open 5-10 per day from the base of the inflorescence upwards over a period of 10-20 days.

***9.4.7 Calyx*:** Calyx lobes 3, erect, ovate triangular, fleshy, much smaller than corolla.

***9.4.8 Corolla*:** Petals 3, free, erect, oblong-linear, rather acute, white at the base, for the rest blue-violet or pale lilac, 1.5-2.5 cm long.

***9.4.9 Androecium*:** Stamens 6, filament filiform, anthers bilobed.

***9.4.10 Gynoecium*:** Ovary inferior, tricarpellary, syncarpous, 10-15 ovules in each of the 3 locules, a narrow style with three lobed stigmas.

***9.4.11 Fruit*:** Multiple fruit (Sorosis), fleshy, hard and juicy when ripe, on the apex with a dense crown of green leaves; normally seedless and ovules abort and only traces of them can be found in mature fruits.

***9.4.12 Seed*:** Small, usually abortive.

9.5 Floral Biology

The flowers open between 6-9 AM which will be fading by evening. The anthesis starts from basal flowers. About 5-10 flowers open daily. By the end of 2-3 weeks after anthesis there will be a complete axis of inflorescence. The anther dehiscence starts soon after anthesis. Open flowers begin to wither in late afternoon. Almost all cultivars produce viable pollens but cannot be self-fertile because of self-incompatibility. Hence, they do not produce seeds after self-fertilization. Cross-pollination between cultivars, cultivars and wild forms leads to normal fertilization. The humming birds are main pollinating agents in South America. In all other parts no such birds, so pineapple fruits are parthenocarpic and seedless.

9.6 Pollination

Due to self-incompatibility, cross-pollination is normal when pollinating agents like humming birds are available. Otherwise parthenocarpic seedless fruits develop.

9.7 Emasculation and Crossing Techniques

Both pollen and ovule are functional but there is self-incompatibility. This is an advantage because it removes tedious process of emasculation. Breeding pineapple includes cross-pollination by hand in the absence of natural cross-pollination. It is unnecessary to emasculate and protect the female parent of a cross, to cover a male parent or to cover or protect the female parent after pollination.

9.8 Breeding Objectives

- To develop high yielding, early maturing cultivars with wide geographical adaptability.
- To develop cultivars suitable for table purpose having medium sized fruits (1-2 kg), cylindrical in shape, sweeter in taste.
- To develop cultivars suitable for canning purpose—bigger fruit size (>2kg), cylindrical in shape, sweeter in taste with high juice content.
- Plant should be hardy, vigorous, have capacity to produce good ratoon crops, leaves should be spineless.
- Fruit stalk should be short and strong.
- There should be flat eyes and small clones.
- To develop cultivars resistant to biotic and abiotic stresses, e.g, heart rot, root rot, wilt, nematodes etc.

9.9 Horticultural Classification of Pineapple

The horticultural classification of pineapple cultivars of Hume and Miller (1904) is currently followed. They divided cultivars of pineapple into three main groups, *viz*, Cayenne, Queen and Spanish. Cayenne group is the most important group. Most of the cultivars in India may be accommodated into any one of the three groups. For example, Kew (Giant Kew) that is grown extensively in India belongs to Cayenne. Queen belongs to Queen group. The cultivars of Cayenne and Spanish groups are used for dual purpose. On the other hand, cultivars of Queen group are grown extensively for fresh-fruit markets, as they are not suitable for canning due to deep eyes.

9.10 Breeding Methods

9.10.1 Selection

Most of the improvement in pineapple has taken place by simple selection of mutant clones within cultivars and by hybridization between cultivars followed by selection of the highly heterozygous progeny. Selection in 'Singapore Spanish' population in Malaysia had led to a new cultivar called 'Masmerah'. Several such selections have been made in different pineapple growing areas.

9.10.2 Hybridization

A cross between Red Spanish and Cayenne has led to the development of a new hybrid PR1-67 in Puerto Rico. Self-fertile somatic mutants obtained from cultivar Cayenne show a loss of vigour on selfing and heterosis on crossing. A hybrid namely H-7 has been produced by crossing Valera Monendi and Kew.

9.10.3 Mutation Breeding

Utilization of mutagenesis in pineapple seems to be quite feasible. However, due to wide natural variation limited attempts have been made for induced mutation. No cultivar has been released through mutation breeding so far.

9.10.4 Biotechnological Approaches

Attempts have been made for rapid multiplication of plants through micro-propagation using different kinds of explants, namely, leaf base, shoot base, excised later buds, meristem tips from crown etc. In the crossing where fertilization fails due to incompatibility, embryo culture technique can help to rescue the hybrid. The genetic transformation of pineapple clones has been attempted with the objective to acquire ability to introduce desirable genes.

9.11 Cultivars

Kew (Giant Kew): Fruits big (1.9 kg), fibreless, juicy with characteristic flavour; leaves long, margin straight and not serrated; eyes broad and shallow (not deep-seated); suitable for canning purpose.

Queen: Fruits small (1.2 kg); flesh sweet, deep yellow and with high flavour; leaves spiny, margin predominantly serrated; quality superior to Kew, eyes deep-seated, not suitable for canning, suitable for table purpose.

Hybrid-7: Fruits 3.0 – 3.8 kg.

Mauritius: Mid-season cultivar, fruits medium (1.5-2.5 kg); flesh red or yellow and fibrous, crown spiny.

Jaldhup: Grown in Assam; sweetness well blended with acidity; with characteristic alcoholic flavour.

Amritha: High yielding hybrid (Kew × Ripley Queen); fruit more than 2 kg, with single, small crown and attractive golden yellow colour, pleasant aroma and cylindrical shape; high TSS and total sugar and low acidity.

Review Questions

Part A

Answer the following questions true or false.

1. Some cultivars of pineapple are used for extraction of fibre. True/False
2. Cultivars of Queen group are suitable for canning. True/False
3. There is self-incompatibility in pineapple. True/False
4. There should be flat eyes in pineapple. True/False
5. The flowers of pineapple open between 9-10 AM. True/False

Part B

Answer the following questions.

1. Why emasculation is not required in pineapple?
2. Why limited attempts have been made for induced mutation in pineapple?
3. Mauritius is mid/late season cultivar.
4. *Ananas comosus* is diploid/triploid.
5. — is considered as ancestor of cultivated pineapple.

Part C

Write a brief note on each of the following.

1. Breeding objectives of pineapple.
2. Breeding methods applied in pineapple.
3. Cultivars of pineapple.

10

Breeding of Litchi

10.1 Introduction

Litchi is a delicious fruit and generally consumed as a table fruit. It is one of the most popular fruits in India. It is also recognized as "Queen of Fruits". It is a subtropical evergreen fruit crop. Its climatic requirements are highly specific. Thus, its commercial cultivation is restricted to only a few subtropical countries of the world. The major litchi producing countries are China, India, Taiwan, Thailand and Vietnam. In India, the important litchi growing states are Bihar, Jharkhand, West Bengal, Tripura, Uttar Pradesh, Uttarakhand, Chhattisgarh, Punjab and Himachal Pradesh. Litchi makes an excellent canned fruit and highly flavoured squash is prepared from inferior fruits.

10.2 Origin and Distribution

Litchi is indigenous to Southern China. It reached the West Indies in 1775, South Africa in 1869, the Hawaii Island by 1873 and Florida in 1883. Other countries, where it reached include Vietnam, Indonesia, Japan, Formosa, Australia, New Zealand, Brazil etc. Litchi reached India through Burma and was first introduced in Bengal during the end of the 17th enntury.

10.3 Taxonomy and Classification

The taxonomic position of litchi as per Hutchinson's system is:

Phylum : Angiospermae

Sub-phylum : Dicotyledones

Division : Lignosae

Order : Sapindales

Family : Sapindaceae

Genus : *Litchi*

Species : *chinensis*

Binomial name : *Litchi chinensis* Sonn.

The *Litchi chinensis* has three sub-species:

Litchi chinensis ssp. *chinensis*

Litchi chinensis ssp. *philippinensis*

Litchi chinensis ssp. *javensis*

The subspecies *philippinensis* is found growing wild in the Philippines. The fruits are distinguished from cultivated litchi by their long and oval shape with long and thorny protuberances and inedible flesh that partially covers the seeds.

The fruit of sub-species *javensis* from the Malay Peninsula and Indonesia produces an aril thinner than that of the cultivated litchi. Neither of these two sub-species is grown commercially. The sub-species *chinensis* is the litchi of commerce.

The chromosome number is 2n=2x= 30.

10.4 Morphology

10.4.1 Habit: Tree medium to large, much branched, round topped, evergreen, height 14 m or sometimes even more.

10.4.2 Roots: Taproot system, symbiosis of *Mycorrhiza* fungus on roots.

***10.4.3 Stem*:** Much branched.

***10.4.4 Leaf*:** Petiolate, exstipulate, compound, mostly paripinnate, with three pairs of leaflets; alternate phyllotaxy, leathery, glossy green on upper side, greyish green on lower side, lanceolate to oblong lanceolate, entire or slightly wavy margin, apex acute or acuminate.

10.4.5 Inflorescence: Consists of several multiple branched panicles, developed from terminal and axillary buds.

10.4.6 Flower: Flowers borne on the axil of the 3rd order, rarely on 4th and occasionally on the main axis in the form of inflorescence; small, apetalous, male, pseudo-hermaphrodite (functional male) and hermaphrodite.

10.4.7 Fruit and Seed: The mature litchi fruit is single seeded nut that usually develops in bunch. The fruit pericarp is papillate and turns pinkish red when fruit is ripe. The aril is an edible portion of fruit which is an outgrowth of the outer cell layer of the seed coat (outer integument). The seeds are dark brown in colour.

10.5 Floral Biology

Anthesis and pollen dehiscence continue throughout the day and night with peak opening in the morning between 6.00 to 9.00 AM. As soon as the stigma starts dividing into lobes, it becomes receptive and remains so up to three days after anthesis.

10.6 Pollination

Litchi requires cross-pollination, carried out primarily by bees.

10.7 Emasculation and Crossing Techniques

Although self-sterility has been reported in litchi, bagging of individual panicles is necessary to avoid out-crossing. To emasculate flowers, scissors are used to remove excess flowers. Only 10-15 flowers are retained for effective crossing. Flowers can be pollinated on the same day or the next day. By dipping a small brush into vials or petriplates, pollens are placed on the stigma. Holding the dehisced anthers by their filaments using forceps can also be adopted to transfer pollen. After pollination, flowers are bagged. Fertilization occurs after a period of several days.

10.8 Breeding Objectives

- To develop precocious, prolific and regular bearer cultivars.
- To breed cultivars that have large size fruits with greater number of fruits per panicle.
- To develop cultivars with alternative skin colour, high aril content, small seed size, high sugar and pleasant aroma.
- To develop cultivars with good keeping quality of fruits.
- To breed cultivars that are resistant to biotic (pests and diseases) and abiotic (fruit cracking) stresses.
- Dwarf stature.
- To develop cultivars with wider agro-ecological adaptability.

10.9 Breeding Methods

10.9.1 Introduction

From China which is the home of litchi, there is much scope to introduce some cracking and pest resistant strains/genotypes.

10.9.2 Selection

Most of the cultivars have been developed through selection. For example, Groft, Brewster, Saharanpur Selection, Calcuttia, Bedana, Dehradun, Rose Scented, Early Large Red, Late Seedless, Swarna Roopa etc.

10.9.3 Hybridization

Intervarietal hybridization resulted in development of Sabour Madhu (H-105), Sabour Priya (H-73) and Sabour Litchi-1. Intergeneric hybridization between litchi and Longan was also attempted, but it didn't produce improved hybrid.

10.10 Cultivars

Swarna Roopa: Local seedling selection, fruit round. TSS 19 %, pulp content 76.62 %, fruit less susceptible to cracking.

Sabour Madhu: Developed from cross Purbi × Bedana, late maturing, medium fruit size, pulp 68.6%, TSS 21 %.

Sabour Priya: Evolved from the cross Purbi × Bedana, fruit quality better than Purbi with respect to aril content and TSS.

Saharanpur Selection: Chance seedling selection, late maturing, TSS 20%, average fruit weight 17 g, less fruit cracking.

Calcuttia: Comparatively more successful in hot and dry areas, heavy bearer and less susceptible to sunburn and cracking.

Rose Scented: Having a distinct rose scented fruit of heart shaped, moderately susceptible to sunburn and cracking.

Seedless Late: Fruits have shrivelled seeds, ripening of fruits starts in 3rd week of June, fruit weight 29 g and TSS 18%.

China: Excellent cultivars grown in West Bengal; plant semi-dwarf, TSS 18%; fruit less susceptible to sunburn and cracking.

Bombaia: Important cultivar of West Bengal, vigorous tree, TSS 17%, suitable for canning.

Early Large Red: Fruit slightly more than 3.4 cm long, usually oblique heart shaped crimson to carmine red in colour with green interspaces. Skin very rough, firm and leathery adhering slightly to flesh; flesh greyish-white, firm, sweet and flavoured, moderate bearer and early maturing.

Early Bedana (Early Seedless): Popular early cultivar in Bihar, UP, Uttarakhand, Punjab and Bangladesh; tree medium, canopy attaining average height of 5.0 m and spread of 6.2 m, regular bearer and medium yielder (50-60 kg/tree), fruit medium sized (15-18 g), oval or heart shaped with deep red skin at maturity.

CHES-2: Developed as clonal selection from Bombaia, late maturing, inside canopy bearing habit, fruit free from sunburn and cracking, fruit deep red, conical shaped and appear in a cluster of about 15-20, average fruit weight 21.3 g containing 3.8 g seed and 16.1 g pulp, TSS 19.8 ^{0} Brix and acidity 0.20 %, Skin : Pulp : Seed ratio :: 18.0: 66.7: 15.3.

Shahi: Released in 2001, fruits attractive deep red with high fragrant pulp (65%) and TSS 20.4 ^{0} Brix; earliest maturing by second week of May; recommended for Bihar, Jharkhand, Uttarakhand, West Bengal and Odisha. Average yield 100-120 kg/tree.

Ambika Litchi-1: Selection from local material; less fruit cracking (6%) as against 28.62% in Shahi. Sugar and TSS at desirable level; tolerant to leaf miner, grey weevil, stink bug, leaf folder and bark eating caterpillar.

Indira Litchi-2: Minimum fruit cracking (5.91%) during high temperature; recommended ecology upland and upland bunded irrigated condition in Northern Hill Zone of Chhattisgarh; TSS 17.9%, pulp 72.2%, average fruit weight 16.69 g and average yield 42.62 kg/tree.

Sabour Litchi: Released in 2015, hybrid, matures in 1st week of June; average fruit weight 23 g; no fruit cracking problem.

Review Questions

Part A

Answer the following questions true or false.

1. Litchi is indigenous to Northern China. True/False
2. The chromosome number of litchi is 2n=2x= 30. True/False
3. The fruit of sub-species *javensis* from Malaya Peninsula and Indonesia produces an aril thicker than that of cultivated litchi. True/False
4. Rose Scented cultivar is developed through hybridization. True/False
5. Bombaia is an important cultivar of litchi in West Bengal. True/False

Part B

Answer the following questions.

1. Litchi is recognized as Queen/King of Fruits.
2. *Litchi chinensis* has three/four sub-species.
3. Name two litchi growing states in India.
4. Sabour Litchi-1 is developed through —.
5. Cite one breeding objective of litchi.

Part C

Write a brief note on each of the following.

1. Methods of breeding in litchi.
2. Cultivars of litchi.
3. Breeding objectives of litchi.

11

Breeding of Ber

11.1 Introduction

Ber (also called Indian Jujube, Indian Plum, Chinese Apple and Masau) is an important minor fruit of India. It is most widely cultivated in Punjab, Haryana, Rajasthan, Uttar Pradesh, Madhya Pradesh, Bihar, Gujarat, Maharashtra etc. Fruits are mostly eaten fresh but other forms, such as dried, candied, squash etc. can be prepared from ber. Stem, bark, root and leaves have some medicinal values. Leaves of ber are used as fodder in dry regions.

11.2 Origin and Distribution

Ziziphus mauritiana Lamk. is supposed to be native of India while *Z. jujuba* Mill. is native to China. Liu and Cheng (1995) reported that Indo-Malaysia region is the centre of both evolution and distribution of the *Ziziphus*. However, De Candolle (1886) stated that Myanmar (Burma) and India are the home of the ber. Maximum variability of ber is found in Rajasthan, Gujarat, Haryana, Punjab and Madhya Pradesh. The central part of the country is supposed to be the richest area in wild ber germplasm.

11.3 Taxonomy and Classification

The taxonomic position of ber as per Hutchinson system is:

Phylum : Angiospermae

Sub-phylum : Dicotyledones

Division : Lignosae

Order : Rhamnales

Family : Rhamnaceae

Genus : *Ziziphus*

Species : *mauritiana*

Binomial name: *Ziziphus mauritiana* Lamk.

Most of the cultivated ber cultivars are tetraploids (2n=4x=48)

11.4 Morphology

11.4.1 Habit: Thorny shrub or small tree up to 12 m in height.

11.4.2 Root: Tap root system.

11.4.3 Stem: Erect, branched, woody.

11.4.4 Leaf: Alternate, simple, 3-nerved, minutely denticulate, obtuse, broadly oval to rounded-elliptical, up to 8 cm long and 5 cm broad, densely tomentose underside, stipular thorns present.

11.4.5 Inflorescence: Axillary cyme.

11.4.6 Flower: Small, greenish-cream, bisexual, actinomorphic, pentamerous, up to 0.8 cm in size, acrid-scented.

11.4.7 Calyx: 5-lobed, shortly tubular, lobes valvate.

11.4.8 Corolla: Petals 5, smaller than calyx lobes.

11.4.9 Androecium: Stamens 5, opposite to petals and arising outside the margin of disk, anthers bilocular and opening lengthwise.

11.4.10 Gynoecium: Ovary superior to half inferior, bilocular, ovule in each locule on basal placentas, style is shorter than stamens, stigma bilobed.

11.4.11 Fruit: Drupe, persistent lower part of calyx, ellipsoid to sub-globose, greenish yellow to golden yellow, 2.2-5 cm long and 2-3.5 cm broad; edible portion epicarp and mesocarp.

11.4.12 Seed: Endospermous (albuminous) with stony seed coat.

11.5 Floral Biology

Anthesis is observed to be cultivar dependent. Anthesis in Seb and Sanaur-2 takes place between 7.30 AM and 8.00 AM while in Gola, Katha and Umran at 1-2 PM. In most of the cultivars, the dehiscence of anthers starts just after anthesis and is completed within 4-5 hours. Stigma is most receptive on the day of anthesis.

11.6 Pollination

Ber does not set any fruit by self-pollination, thereby shows self-incompatibility. Prominent pollinating insect agents are honey bees, house flies and yellow wasp. Generally, cross-pollination is rule in ber.

11.7 Breeding Objectives

- To develop early maturing cultivars.
- To evolve drought resistant cultivars.
- To develop superior quality and high yielding cultivars.
- To develop cultivars resistant to biotic stresses, e.g., fruit fly and powdery mildew.
- To breed dwarf stature cultivars suitable for high density planting.
- To develop less thorny cultivars.

11.8 Breeding Methods

11.8.1 Selection

Most of the common cultivars are the result of selection made by local people in different regions. The promising cultivars under commercial cultivation are Gola, Seb, Banarasi Karaka, Banarasi Peondi, ZG 1, Sanaur-1,2,3,4, Katha, Umran, Mundia etc.

Goma Kirti is clonal selection from Umran. CIAH Sel-1 is developed through selection from local material collected from Bhusawar area of Rajasthan.

11.8.2 Hybridization

At HAU, Hisar, 72 hybrids were raised and are under evaluation. Hybridization was also initiated at Jodhpur by making cross of Katha × Seb, and Seb × Tikadi. At CIAH, Bikaner, CIAH Hybrid-1 (Seb × Katha) was found promising.

11.9 Cultivars

Goma Kirti: Early maturing, matures 3 weeks earlier than Umran, high yielding (25.8-38.2% increase over control); physiological loss in weight 12% against 27% in Gola; shelf-life 6 days against 3 days in case of Gola; fetches good price in distant market.

Thar Sevika: Released by CIAH, Bikaner in 2006, developed by hybridization (Seb × Katha), early maturing, average fruit yield 30-32 kg/tree.

Thar Bhubharaj: Released by CIAH, Bikaner in 2007; selection from local material; average yield 30-36 kg/tree, fruits very juicy, sweet, TSS 22-23%.

Review Questions

Part A

Answer the following questions true or false.

1. *Ziziphus mauritiana* Lamk. is supposed to be native of China. True/False
2. The seed of ber is albuminous. True/False
3. The flower of ber is unisexual. True/False
4. Goma Kirti is clonal selection from Umran. True/False
5. Generally, self-pollination is rule in ber. True/False

Part B

Write a brief note on each of the following.

1. Cultivars of ber.
2. Breeding objectives of ber.
3. Selection and hybridization in ber.

- To develop cultivars suitable for export.
- To evolve coloured cultivars based on market demand.
- To breed cultivars with high percentage of female flower with good bearing.
- To breed cultivars having high yield with good quality fruits.
- To develop cultivars suitable for processing purpose.
- To breed cultivars with less fibre content.
- To develop cultivars resistant against biotic and abiotic stresses, e.g., rust disease, frost, drought etc.

12.7 Breeding Methods

Selection

In India, breeding in aonla has so far been limited only to selection technique through which several high yielding cultivars were identified and released for commercial cultivation.

Examples: Agra Bold, Banarasi, Goma Aishwarya, Kanchan (NA-4), NA-9, NA-10, Amrit (NA-6), Francis, Krishna (NA-5), Neelum (NA-7), Balwant, (NA-10) Chakaiya, BRS-1, Anand-1, Anand-2, Anand-3 and Lakshmi-52.

Other methods of breeding are not yet attempted in aonla.

12.8 Cultivars

Krishna: Chance seedling of Banarasi; bears moderately, fruits medium to large (44.6 g), conical at apex and apricot yellow in colour; one female flowers/ branchlet; fibre- 1.4%, vitamin C 783 mg/100 g.

NA-9: Chance seedling of Banarasi; bears moderately, 1.2 female flowers/ branchlet; fruits large (50 g), flattened, smooth, yellowish skin, 6-8 strips; fibre low (0.9%); ascorbic acid very high (881 mg/100 g).

Balwant: Chance seedling of Banarasi; bears profusely; fruits attractive, medium to large (41.5 g), shape flattened, round, skin rough, yellowish green with pink tinge.

Neelum: Selection from Francis; semi-erect growth, profuse bearing due to high female flowers (5 per branchlet); precocious; fruits medium to large (47.5 g), shape flattened, oblong with conical apex, flesh almost fibreless, soft with moderate keeping quality.

Kanchan: Chance seedling of Chakaiya; spreading tree growth; bears profusely due high number of female flowers (4-6/branchlet); fruits small to medium (30-32 g), flattened, oblong, skin smooth.

Amrit: Chance seedling of Chakaiya; fruits medium to large (35-37 g) with bright appearance; stripes 6- distinct and thin; lowest fibre content (0.8%); profuse bearing due to high number of female flowers (3.8/branchlet).

BRS-1: Selection from local cultivar Thimbam grown in Tamil Nadu; yield potential very high (155 kg/tree); fruits have more flesh with less phenolic content; ascorbic acid content very high.

Laxmi 52: Chance seedling of Francis; semi-erect growth, branches do not drop like its parent Francis; fruit large (40-60 g) with 6 ridges; fruit colour light pink that disappears on full development, free from necrosis; yielding potential 2-2.5 q/tree (after 10 years).

Review Questions

Part A

Answer the following questions true or false.

1. Aonla is indigenous to tropical South Eastern Asia. True/False
2. Laxmi 52 is chance seedling of Kanchan. True/False
3. Aonla is highly cross-pollinated. True/False
4. Selection is only method attempted in aonla in India. True/False
5. Aonla belongs to family Asteraceae. True/False

Part B

Write a brief note on each of the following.

1. Cultivars of aonla.
2. Floral biology of aonla.
3. Breeding objectives of aonla.

13

Breeding of Pomegranate

13.1 Introduction

Pomegranate (*Punica granatum* L.), an economically important fruit plant species, belongs to the family Punicaceae. It is an important fruit crop of arid and semiarid regions of the world and has been believed to be the symbol of fertility and abundance. Pomegranate is a highly remunerative crop for replacing subsistence farming and alleviating poverty. The plant is drought tolerant, winter hardy and can thrive well under desert condition. It is a good source of protein, carbohydrate, minerals, antioxidants, vitamins A, B and C, and has also been used in controlling diarrhoea, hyperacidity, tuberculosis, leprosy, abdominal pain and fever. Pomegranate juice contains antioxidants such as soluble polyphenols, tannins, anthocyanins and can be used in the treatment of cancer and chronic inflammation. It has enjoyed a reputation for its healthy dietetic and medicinal properties. The fruit juice is considered to be useful for patients suffering from leprosy. The bark and the rind of the fruit are commonly used against dysentery and diarrhoea. In recent years the area under this crop has increased substantially, mainly because of versatility, adaptability, drought resistance, low maintenance costs, and steady and high yield. The fruit has a good consumer preference for its attractive, juicy, sweet and acidic arils. There is a great demand for growing quality fruits both for fresh use and processing into juice, syrup, squash, wine and anardana. The pomegranate is capable of growing in a wide range of soils as long as it has good drainage. The pomegranate is tolerant to drought and it prefers very warm temperatures, which improve the flavour of the fruit.

Now, it is grown commercially in Egypt, Iran, Spain, Morocco, Syria, Afghanistan, Turkmenistan, Pakistan (Baluchistan) and India. It is also grown to some extent in Myanmar, China, Japan and USA (California). In India pomegranate is mainly cultivated in Maharashtra, Karnataka, Andhra Pradesh, Uttar Pradesh, Gujarat and Tamil Nadu. Pomegranate cultivation is one of the most remunerative farming enterprises in India. Several types of pomegranate cultivars are cultivated in India and are distinguished by the shape of the fruit, the colour and thickness of rind, and the taste and colour of the seeds. Being a cross-pollinated crop, a lot of variability exists in seedling populations which can be utilized in further improvement programme. Though pomegranate is considered as tolerant to a wide range of climatic conditions, the high variability of new growing environments is a great challenge to the commercial production of high-quality fruit and yield. Therefore, breeding for new pomegranate cultivars that are better adapted to variable climatic conditions becomes an important issue for pomegranate production. Most pomegranate cultivars grown today are the result of human selection from naturally occurring cultivars. Until recent years, pomegranates were selected according to the demands of local consumers and not for export. Therefore, the main cultivars found today reflect the local priorities of each country or region. Examples are the traditional Indian and Spanish cultivars that are characterized by their soft seeds and low-acid taste. Increased world demand and economic importance of pomegranate exports should significantly influence pomegranate selection criteria, which will have an increasing role in pomegranate breeding. Traditionally pomegranates were selected on the basis of their juice content, fruit size, colours, yield, and taste. The preferences of skin or aril colour and taste are not identical in all countries. Export considerations raise the importance of ripening time, skin and aril colour, taste, and health benefits as prioritized by consumers in the end markets.

A large number of diverse seedling forms have evolved in nature in various regions of the world over the years. Very few cultivars have been developed by systematic breeding programmes despite great opportunities exists for developing cultivars to meet various objectives. Breeding of pomegranates in India was done for disease resistance, low acidity and high fruit quality under hot arid environment, fruit yield, juice production, aril, fruit weight, flesh colour, seed size, juice content, TSS and seed softness.

13.2 Origin and Distribution

Currently wild pomegranates are grown in central Asia from Iran and Turkmenistan to northern India. Pomegranate is considered as native to these regions. Pomegranates were introduced throughout the Mediterranean region, to the rest of Asia, to North Africa and to Europe. Pomegranate is cultivated today throughout the world in subtropical and tropical areas in many different microclimatic zones. Commercial orchards of pomegranate trees are now grown in the Mediterranean basin (North Africa, Egypt, Israel, Syria, Lebanon, Turkey, Greece, Cyprus, Italy, France, Spain, and Portugal) and in Asia (Iran, Iraq, India, China, Afghanistan, Bangladesh, Myanmar, Vietnam, Thailand; and in the former Soviet republics: Kazakhstan, Turkmenistan, Tajikistan,

Kirgizstan, Armenia, and Georgia). In the New World, pomegranates are grown in the United States and Chile.

13.3 Taxonomy and Classification

According to the phylogenetic system of classification of Hutchinson (1959), pomegranate is classified as:

Phylum - Angiospermae

Sub-phylum -Dicotyledones

Division - Lignosae

Order - Myrtales

Family- Punicaceae

Genus - *Punica* L.

Species - *granatum*

Binomial name - *Punica granatum* L.

There are two species under the genus *Punica*, *Punica granatum* and *Punica protopunica*. The taxonomy of pomegranate remains poorly understood, in spite of numerous previous phylogenetic studies. It belongs to the monogeneric family Punicaceae.

13.4 Morphology

13.4.1 Habit

Pomegranate is a shrub that naturally tends to develop multiple trunks and has a bushy appearance. When domesticated, it is grown as a small tree that grows up to 5 m. Under natural conditions, it can sometimes grow up to more than 7 m. There are also dwarf cultivars that do not exceed 1.5 m. Most of the pomegranate cultivars are deciduous trees. However, there are several evergreen pomegranates in India. Pomegranates are also long-lived. The vigour of a pomegranate declines after about 15 years.

13.4.2 Root

Shallow root system with most of being less than 60 cm deep and very rarely below 90 cm.

13.4.3 Stem

The stem is covered by a red-brown bark which later becomes grey. The branches are stiff, angular and often spiny. There is a strong tendency to sucker from the base.

13.4.4 Leaf

Leaves relatively small, slender oblong, almost sessile. The leathery leaves are narrow and lance shaped. Young leaves tend to have a reddish colour that turns green when the leaf matures. In cultivars with young pink-purple bark, this colour appears also on the sheath and the petiole, on the lower part of the central vein, and in the leaf margins. The leaves are exstipulate, opposed and pairs alternately crossing at right angles.

13.4.5 Inflorescence

Dichasial cyme and due to heavy drop of secondary and tertiary buds, they appear to be solitary in clusters.

13.4.6 Flower

Flowers are large (> 2.5 cm), attractive, red, scarlet, white or variegated, funnel shaped, borne terminally on clusters of 1 to 5. The flowers may be solitary or grouped in two's and three's at the end of the branches. In evergreen cultivars in southern India, flowering season is observed in three periods: June, October, and March. In the early balloon stage, the flower resembles a small pear with a greenish colour on its basal part and reddish colour on its apex or entirely dark red. *Pomegranate develops one of the two types of flowers normally: (1)* ***hermaphrodite (bisexual) flowers (vase-shaped)*** *that are complete, actinomorphic, epigynous, and (2)* ***male flowers (bell-shaped)*** *that are incomplete, actinomorphic.* Both types have several hundred stamens. Male flowers are significantly smaller than bisexual flowers. The bell-shaped flower has a poorly developed or no pistil and atrophied ovaries containing few ovules and is infertile. Therefore, the bell-shaped flower is referred to as a male flower and will drop without fruit set. The vase-shaped flower is fertile with a normal ovary capable of developing fruit. The stigma of the hermaphrodite flower is at the anther's height or emerging above them. This position allows for self-pollination as well as pollination by insects. The factor that determines the fruit set capacity is the number of vase-shaped flowers. Therefore, cultivars with higher vase-shaped to bell-shaped flower ratio will have a higher fruit yield potential.

13.4.7 Calyx

The sepals (lobes), 5 to 8 fused at their base, form a red fleshy vase shape. The calyx is adnate to the ovary. It is persistent and will not drop with fruit set but will stay as an integral part of the fruit as it matures, generating a fruit crowned with a prominent calyx. The aestivation is valvate.

13.4.8 Corolla

The corolla has 5 to 8 lustrous petals, the number usually equals to the number of sepals. The petals, which alternate with the sepals, are separated and have a pink orange to orange red colour depending on the cultivar. The petals are obovate, very delicate, and slightly wrinkled. The aestivation is imbricate.

13.4.9 Androecium

The multiple long stamens are inserted into the calyx walls in a circle. There are more than 300 stamens per flower. They have an orange red filament and yellow bilocular anthers that remain attached to the prominent calyx. The time taken by the flowers to complete anthesis is 3 to 5 hours. Pollen from both bisexual and male flower types is similar in size (20 mm). However, male flowers have significantly more stamens per flower than bisexual flowers. Pollen viability in both bisexual and male flowers is relatively high.

13.4.10 Gynoecium

The carpels vary in number but are usually eight superimposed in two whorls. They form a syncarpic ovary and are arranged in two layers. Nectaries are located between the stamens and the ovary base. Ovary is inferior and embedded in thalamus. Style is yellow and fleshy with greenish stigma.

13.4.11 Fruit

Fruits are irregularly rounded with bright red, leathery rind and a prominent calyx. The tough, leathery skin or rind is typically yellow with light or deep pink or rich red. The interior is separated by membranous walls and white, spongy, bitter tissue into compartments packed with sacs filled with sweetly acid, juicy, red, pink or whitish pulp or aril. In each sac there is one angular, soft or hard seed. High temperatures are essential during the fruiting period to get the best flavour. The rind encloses membranous, white tissue (endocarp) which in turn encloses the arils and seeds. The juice is contained in the arils. The pomegranate may begin to bear in 1 year after planting out, but 2.5 to 3 years is more common. Under suitable conditions the fruit mature after 5 to 7 months after bloom. The fruit is botanically known as 'balusta'.

13.4.12 Seed

Inside the fruit there are locules separated by a thin, bitter membrane. Within the locules there are many small jewel-like objects present called as arils. The arils contain sour to sweet flavoured juice and a seed.

13.5 Floral Biology

Flowering in pomegranate is characterized as having both hermaphroditic (bisexual) flowers and functionally male flowers on the same plant, a condition referred to as **andromonoecy**. Inflorescence of the pomegranate has been reported to be the cyme. Nath and Randhwa (1959 a) observed the inflorescence to be dichasial cyme and due to heavy drop of secondary and tertiary buds, they appear to be solitary in clusters. In the evergreen pomegranate cultivars, the flower buds of the spring flush are borne on mature wood of one-year old shoot whereas the flower which appear during July-August are borne on the current season growth. In deciduous cultivars, flowers are borne on the current season growth between July and August. The hermaphroditic flowers have well-formed female (stigma, style, ovary) and male (filaments and anthers) parts and have been referred to as fertile, vase-shaped and bisexual flowers. Because the hermaphroditic flowers are the type that set fruit, they are commonly referred to as "female" flowers. The male flowers produce well-developed male parts, but on closer examination of the pistil contain reduced female parts. Thus, their role is more accurately depicted as **functionally male flowers** (i.e., flowers are not strictly male), but rather have degenerated female parts. Male flowers typically drop and fail to set fruit. Chaudhari and Desai (1993) classified pomegranate flowers into three types: male, hermaphroditic, and intermediate. Ovary of male flower is rudimentary whereas that of intermediate flowers is degenerating type. If fruit set takes place in such flowers, they may drop before reaching to maturity, even if some fruits reach maturity that become misshaped. **Heterostyly** is also observed in pomegranate wherein the hermaphrodite flowers are **pin eyed** and male flowers are **thrum eyed**. Hermaphrodite flowers are usually homostylous or pin eyed (i.e. the stigmas are on the same level or higher than the anthers) and male flowers are thrum eyed (i.e. the stigmas are beneath the level of the anthers). The hermaphrodite flowers resulted in fruit-set and fruiting; staminate flowers are meant for pollination, intermediate flowers may or may not set fruits. The hermaphrodite flowers have a higher fruit set than intermediate flower. The floral biology varies with the cultivar, location and season. Anthesis in pomegranate begins at 8.00 A.M. and is completed by 4.00 P.M. with peak at 2.00 P.M. The time and duration of dehiscence of anthers vary with the cultivar and no general sequence is found at the time of anthesis. The stigma attains receptivity one day before anthesis and remains in receptive condition up to the second day after anthesis.

13.6 Pollination

The pomegranate is self-pollinated as well as cross-pollinated by insects. Cross-pollination increases the fruit set. Male flowers drop and generally fail to set and fruits develop exclusively from bisexual flowers. The degree of fruit set by self-pollination varies among different pomegranate cultivars. In hermaphrodite flowers, 6% to 20% of pollen may be infertile; in male flowers, 14% to 28% are infertile. The size and fertility of the pollen vary with the cultivar and season (Morton 1987).

13.7 Emasculation and Crossing Techniques

Good fruit set is observed when crossing is done during winter season due to higher percentage of viable pollen during low temperature and high humidity conditions. Hybridization is carried out through emasculation of flowers shortly before anthesis using forceps followed by pollination immediately. Alternatively, the flower buds to be opened in the next morning are selected for hybridization. Stamens are taken out by inserting forceps through petals. This is followed by bagging and tagging. Desired anthers in yellow stage are collected and cut open with the help of the needle and then pollens were transferred with the help of a brush to the receptive stigma of the emasculated bud in the next morning. Bagging and labelling is done. Hybridization is done in the morning hours before 11 a.m. Fruits require about 120-140 days for maturity. Crossed fruits are harvested and the seeds are extracted. They are sown in plastic trays. Seeds germinate within 10 days of sowing. The seedlings are transplanted to polythene bags and subsequently to field for further evaluation.

13.8 Genetics and Cytogenetics

The cytology of a crop is useful in planning breeding programme. The cultivated species of pomegranate (*Punica granatum*) is reported to contain 2n=16, 18 chromosomes. Some of the cultivars have chromosome number 2n=16 while the Double flower has 2n=18.The chromosome number in cultivars Dholka, Ganesh, Kandhari, Muscat White and Patial is found to be 2n=16. The chromosome number in Vellodu and Kashmiri cultivars is found to be 2n=18 (Nath and Randhawa,1959b). Further, *Punica granatum* has been classified into two sub-species *viz. Chlorocarpa* and *Porphyrocarpa.* Another species of pomegranate is *P. nana* which is having double flowered habit and generally it is grown as hedge. The pomegranate crop exhibits a long reproductive cycle, heterozygosity and difficulty in raising a massive population due to which studies on genetics of various traits are limited. The hard-seeded nature and red and pink aril colour are dominant to soft seediness and white aril colour. Fruit acidity is monogenically controlled and that sourness or higher acidity is dominant over sweet or low acidity. Very little is known on the heritability of desirable traits in pomegranates. Resistance to bacterial leaf blight is controlled by recessive genes.

13.9 Bottlenecks of Fruit Breeding in Pomegranate

Pomegranate cultivation is still faced with many problems and methods must be developed for cultivar identification and improvement and genetic resource management. Since there is increasing demand of pomegranate over the world for its superior pharmacological and therapeutic properties, there is a need to initiate and intensify well-planned breeding programmes to meet the need of local and international consumers, processors, growers and exporters. Pomegranate generally shows relatively long pre-bearing period, heterozygosity and huge populations are required for screening for the particular traits. Besides, the following are some of the problems associated with pomegranate breeding:

1. Non-availability of bacterial blight resistant cultivars.
2. Non-availability of donor parent with bacterial blight resistance.
3. Difficulties in identification of tolerant cultivars for wilt disease, the cultivar has to be screened against the fungus *Ceratocystis fimbriata*, shot hole borer (*Xyleborus* sp.) and root knot nematode (*Meloidogyne incognita*) as they are individually or collectively responsible in causing wilt.
4. Susceptibility of the commercial cultivars to pests and diseases: Most of the cultivars are susceptible to insect-pests (thrips, fruit borer, fruit sucking moth, etc.) and diseases (bacterial blight, wilt).
5. Susceptibility of cultivars to fruit cracking.

13.10 Breeding Objectives

Since pomegranate has become an economically important fruit crop and is being exported, breeders should consider criteria of international market like fruit size, fruit shape, skin colour, aril colour, easily separable bold arils, seed size, sugar : acid blend, shelf-life, good post-harvest quality. Following objectives are to be taken in to consideration for breeding in pomegranate:

- To develop easily manageable upright growth habit of the tree.
- To develop thornlessness in the twigs, a desirable character as it helps in cultural management of the tree.
- To develop short duration cultivars (130-140 days) like Mridula, and Phule Arakta.
- To develop anardana type cultivars with dark red and bold arils, high acidity, bigger fruit size, with high yield.
- To develop fruit borer (*Virachola isocrates*), fruit rot (*Phomopsis* sp.) and soil borne fungi especially *Fusarium* resistant cultivars.
- To develop cultivars with better fruit quality with respect to fruit colour, soft seeds and red coloured seeds.
- To develop cultivars free from fruit cracking and blackening of arils.
- Identification and development of suitable cultivars for cold arid region.
- To develop cultivars with longer storage life.

13.11 Breeding Methods

13.11.1 Introduction

Introduction is the method by which cultivars are introduced from the place of cultivation to another place where it was not cultivated before. Introductions like Gulsha Red Pink, Appuli, Gulsha Red, Lupania, Bedana Sedana, Kandhari and Khog are being used in breeding programmes to induce dark red aril pigment in the popular cultivars (Pareek and Sharma, 1993). In recent years, some commercial cultivars including soft seeds, dark red grained types, *viz.* Wonderful from USA; Males, Be Hastah, Alah, Agha Mohammad Ali, Post Sephid, Sirin from Iran and Rannyiz G-1-8-23, Rannyiz G-1-3-34, Cherenyj G-1-8-7 and few cultivars from Tunisia have been introduced (Singh and Rana, 1993). At Hisar, cultivar Shirin Anar and Russian seedling are found resistant to bacterial leaf spot. The important introduced cultivars of pomegranate are Wonderful (originated in Florida) and Granada (originated in California as a bud mutation of wonderful).

13.11.2 Selection

Many pomegranate types cultivated in India are of seedling selection origin. They offer a wide range of variability with respect to shape and size of fruits, mellowness of seeds, aril colour, rind colour, sweetness and acidity. Due to considerable variability and their adaptability to existing agro-climatic conditions, selection of superior genotypes will be the best approach to get desirable ideotypes. On the basis of yield and physico-chemical characters of fruits, number of cultivars have been recommended for commercial cultivation in different states of India, *viz.*, Ganesh, G-137, P-23, P-26 and Muscat in Maharashtra, Bassein Seedless, Jyothi and Madhugiri in Karnataka, Dholka in Gujarat, Jalore Seedless, Jodhpur Red, and Jodhpuri White in Rajasthan and Kabul Red, Vellodu, Yercaud 1, and Co-1 in Tamil Nadu. Two ornamental types (Japanese Dwarf and Double flower giving red, yellow and white flowers) are planted in the ornamental gardens (Nath and Randhawa, 1959). The cultivar GBG-1 is a selection from open-pollinated population of Alandi in 1932. The name Ganesh was given in 1970. Five Muskat types, namely, P-13, P-16, P-23, P-26 and SK-1 were identified by Naik (1975). Further, P-23 and P-26 were released in 1986 for commercial cultivation by MPKV, Rahuri. At the University of Agricultural Sciences, Bangalore as a result of evaluation of seedling populations raised from Bassein Seedless and Dholka cultivars, GKVK-1 now named as Jyothi was released. At Coimbatore self-seeded selection Co-1 was identified (Khader *et al.*, 1982).

13.11.3 Hybridization

Hybridization is the process of crossing of genetically two different parents. It is done by emasculation and hand pollination. *Breeding in pomegranate is somewhat easier as compared to breeding of quite a lot of other fruits as it bears large flowers making hybridization handy and produces fruits having abundant seeds which generally germinate well and the crop has relatively shorter juvenile phase.* In India, main emphasis in hybridization works is to breed coloured pomegranate cultivars with good size and taste. For development of coloured cultivar, Russian cultivars have been used as these cultivars produce dark red arils often with red fruit skin but have hard seeds and produce small acidic fruits and do not come up well under Indian condition. In order to incorporate blood red colour of Russian types into Ganesh, several crosses were made at Rahuri in 1976.Out of 122 F_1 hybrids, seven had deep red aril colour but the seeds were hard and inferior in taste to Ganesh. A promising line from the F_2 population (No.61) combining desirable quality attributes has been released by the name Mridula (Ganesh × Gulsha Rose Pink). **Two of the main commercial export cultivars from India, Mridula and Bhawa, are the result of a selection from progenies of a cross between 'Ganesh' and the red 'Gul Shah Red' pomegranate cultivar from Russia.** Ganesh itself is an evergreen selection from 'Alandi' that produces a soft-seeded fruit with poor fruit quality. Mridula and Bhawa combine the red skin colour, seed softness, and evergreen habit of growth from their parents.

13.11.4 Mutation Breeding

Use of physical (x- rays) and chemical mutagens (N, N-dimethyl N-nitrosourea) may help in the development of the superior cultivar of soft seeded types (Levin, 1990).

13.12 Important Cultivars

Alandi: Also known as Vidaki. Fruit is medium in size (170 to 190g), round and smooth with nipple at pedicel end and pink in colour. Arils are sweet in taste with light pink in colour.

Amlidana: It was hybrid (F_1) between Ganesh and Nana (dwarf ornamental miniature cultivar) released from IIHR, Bangalore. Fruits are highly acidic. It is recommended for commercial cultivation for anardana. Fruit is round and smooth and pink in colour with reddish tinge.

Bassein Seedless: This cultivar is cultivated to some extent in Karnataka state. Tree is evergreen with spreading habit. The fruit is round and weight varies from 260 to 300g. The outer rind colour is red; the seeds are soft having light pink coloured arils.

Bedana: Fruit medium to large in size, rind brownish or whitish, fleshy testa, pinkish white with sweet juice and soft seeds.

Bhawa: It is developed by MPKV, Rahuri. It is a selection from F_2 progeny of a cross between Ganesh × Gul-e-Shah Red, a Russian cultivar. This is the most popular and ruling cultivar of pomegranate in Maharashtra with maximum area under cultivation. It is suitable for long distance market as it has thick rind, better keeping quality (15-20 days at room temperature). It is tolerant to thrips and mites, it is free from blackening of arils and there is no incidence of fruit cracking. Fruits have cherry red bold aril.

Dholka: Large fruit size (400 to 500g), greenish white rind, fleshy testa, pinkish white or whitish with sweet juice, soft seeds and acidic juice. The cultivar is grown in Dholka area of Gujarat. The growth habit of tree is spreading type with evergreen nature. Fruit is round and smooth and light green in colour with reddish tinge. Arils are sweet in taste with light pink in colour.

Ganesh: Prolific bearer, medium fruit size, soft seeds, sweet in taste. Also known as GBG-1 and selected by Dr. Cheema in 1936. A soft seeded selection from open pollinated seedlings of hard seeded Alandi. Later, in 1970, it was renamed as 'Ganesh'. The fruit matures 140-150 days. Fruit is round and smooth, pinkish yellow to reddish yellow rind colour, having soft seeds and light pink arils, fruits weighing between 225 to 250g, with T.S.S. 16°B with acidity of 0.3%. The arils of this cultivar are sweet in taste with pink colour in winter months and are whitish in warmer months.

G-137: It was released by Mahatme Phule Krishi Vidhyapeeth (MPKV), Rahuri in the year 1989. It is the results of Clonal selection from open-pollinated cultivar Ganesh. G - 137 is superior to Ganesh in respect of skin and aril colour, aril size and TSS (Keskar *et al.*,1990).

Goma Khatta: It is a hybrid between Ganesh × Daru. The hybrid having high acidity (7.3%) and bigger fruit size (137.3g) compared to cv. Amlidana (86.33g). It is superior to Daru. The growth habit of tree is upright type with evergreen and dwarf in nature. Fruit is round and smooth and pink in colour with reddish tinge.

Jalore Seedless: This cultivar is cultivated in Rajasthan and the plants are semi-spreading in habit; this has fruits weighing between 280 to 320g. Fruit is round and smooth. The rind colour is reddish yellow to pinkish yellow. Arils are sweet in taste with light pink in colour. The seeds are very soft.

Jyothi: Also known as GKVK-1 with attractive yellowish red colour fruits. The fruits are medium size with red coloured arils having soft seeds. This is a selection from the seedlings of Bassein Seedless. The growth habit of tree is spreading type with evergreen nature.

Kabul: Large fruit size, rind deep red mixed with pale yellow, thick, fleshy testa dark red, slightly bitter juice. The cultivar is susceptible to fruit borer, leaf spot and fruit spot.

Kandhari: Fruit large in size, rind deep red, fleshy testa, blood red or deep pink with sweet, slightly acidic juice, hard seeds. The cultivar is susceptible to fruit borer, leaf spot and fruit spot. The cultivar is suitable for different parts of Maharashtra and Gujarat states.

Mridula: It is a seedling selection from an open pollinated F_2 population of a cross between Ganesh × Gul-e-Shah Red and released by Mahatma Phule Krishi Vidhyapeeth, (MPKV), Rahuri in the year 1994. This is an early cultivar requiring only 135-140 days. Fruit is round and smooth and pink in colour with reddish tinge. Arils are sweet in taste with light pink in colour. The cultivar is susceptible to fruit borer, leaf spot and fruit spot.

Muskat Red: Fruit small to medium in size, rind somewhat thick, fleshy testa with moderately sweet juice, seeds are semi-hard. The cultivar is susceptible to fruit borer, leaf spot and fruit spot.

Paper Shell: Fruit medium in size, rind thick, fleshy testa, reddish pink with sweet juice and soft seed.

14

Breeding of Apple

14.1 Introduction

Apple is the most widely grown temperate fruit of the world. It is considered as King of temperate fruits. It is rich in carbohydrate, calcium, phosphorous, iron, potassium, vitamin B_6. Apple is rich in antioxidant, flavonoids and dietary fibre. It is recommended to reduce the incidence of dental caries. The phytonutrients and antioxidants may help reduce the risk of developing cancers, hypertensions, diabetes and heart disease. Apple is mainly used for table purpose. The various processed products are prepared from fruit, e.g., juice, jelly, preserve, canned slice, wine, concentrated juice, cider and powder. It is widely cultivated in Afghanistan, Armenia, Azerbaijan, Bhutan, China, India, Indonesia, Iraq, Iran, Israel, Japan, Jordan etc.

14.2 Origin and Distribution

The centre of origin is from Asia Minor to Western Himalaya. Apple was introduced into India by British in the Kullu Valley of Himachal Pradesh as far back as 1865, while the coloured "Delicious" cultivars of apple were introduced to Shimla hills in 1917. The apple cultivar 'Ambri' is considered to be indigenous in Kashmir and had been grown long before Western introductions. It is predominantly growing in temperate zone of India.

14.3 Taxonomy and Classification

The taxonomic position of apple as per Hutchinson system is:

Phylum : Angiospermae

Sub-phylum : Dicotyledones

Division : Lignosae

Order : Rosales

Family : Rosaceae

Genus : *Malus*

Species : *x domestica*

Binomial name : *Malus x domestica* Borkh.

Cultivated apples (*Malus x domestica*) are diploids (2n=2x=34) and few are triploids (2n= 3x=51), and the ancestor is *Malus sylvestris*.

14.4 Morphology

***14.4.1 Habit*:** Tall tree, rarely shrubs, deciduous, rarely evergreen.

***14.4.2 Root*:** Tap root system.

14.4.3 Stem: Woody, erect and branched.

***14.4.4 Leaf*:** Elliptical/oval, bluntly serrated, alternate, stipulate.

***14.4.5 Inflorescence*:** Racemose, umbel.

***14.4.6 Flower*:** Pedicellate, actinomorphic, white to light pink.

***14.4.7 Calyx*:** Free sepals, five.

***14.4.8* Corolla**: Five petals, broadly ovate.

***14.4.9 Androecium*:** Stamen numerous, filaments long, anther two lobed, longitudinally splitting.

***14.4.10 Gynoecium*:** Inferior ovary, pentacarpellary, syncarpous, style-5, long with elevated stigmas.

***14.4.11 Fruit*:** Pome, the fleshy edible portion derives from the hypanthium or floral cup, not the ovary that is why it is considered as the false fruit.

***14.4.12 Seed*:** 5 seed cavities with generally 2 seeds in each. Seeds are relatively small and black and mildly poisonous.

14.5 Floral Biology

The apple trees require chilling hours to break the rest period. It varies from 1000-1600 hours depending on the cultivars. Flowering occurs in early spring. Pollen fertility in most of the cultivars is 100% but it is reduced in some cultivars. Stigma attains receptivity two days prior to anthesis and remains receptive two days after anthesis with peak receptivity on the day of anthesis.

14.6 Pollination

Most apple cultivars require cross-pollination due to self-incompatibility and triploid nature of cultivars. Most of apple cultivars are self-unfruitful and cannot produce fruit if fertilized by their own pollen and thus require some compatible cultivars for cross-pollination and good fruit set. Apple tree produce abundant bloom but fail to set fruit especially under adverse climatic conditions due to lack of pollination. Inadequate fruit set often results from a failure during the pollination period, which is associated with pollen production, transfer and germination, pollen tube development or fertility of the ovule. Sterility and incompatibility are two main causes of unfruitfulness in apple. Self-incompatibility is most common, although cases of cross-incompatibility are also known. The commercial cultivars Red Delicious, Royal Delicious, Top Red, Vance Delicious, Red Chief and Oregon Spur are self-incompatible. Various climatic and edaphic factors are also responsible for poor pollination and fruit set. Low temperature, rainfall and cloudy weather at flowering time adversely affect the bee activity, transfer of pollen to stigma, pollen germination, and ultimately result in poor fruit set. The cross-pollination and fruit set in apple can be improved by planting at least 25 to 33 per cent of pollinizers, placement of 5 to 6 honey bee colonies per hectare, top working of 2-4 shoots of commercial cultivars with pollinizers and placement of bouquets.

14.7 Emasculation and Crossing Techniques

Emasculation is carried out at "balloon stage" and pollination is done after emasculation. As the stigma remains receptive for a number of days, re-pollination can be done a day later to ensure crossing. In case of self-incompatible cultivar there is no need to do the emasculation and a small tree can be caged in a bee-proof structure with a bouquet of open flowers of desired parent and honey bees. One can use pollen parent with a dominant marker in order to separate selfed seeds from the crossed ones. Apple seeds require stratification before germination. Seeds are harvested before maturity of fruits. Seeds are germinated in plastic bags containing filter paper or peat moss. When radicle begins to emerge, the seeds are transferred to pot or trays and kept until the seedlings are of 30-45 cm height. These seedlings are then transferred to nursery or to larger pots before transfer to the field. Characters can be recorded on the plants in the nursery as well.

Phase I: Testing of genotypes whether grafted or not, in single replication.

Phase II: Most promising seedlings are cloned by grafting onto dwarfing rootstocks (4-6 trees per genotype are produced).

Phase III: Multi-location testing comprising either single replication of six trees or one can go for two replications. A total of 10-50 trees are evaluated at multi-sites. Virus indexing (and thermotherapy if required) is done at this stage.

14.8 Breeding Objective

Scion breeding objectives

- To develop early maturing cultivars.
- To develop cultivars with red colour.
- To develop cultivars with high yield.
- To develop cultivars with superior dessert and storage quality.
- To develop cultivars with resistance against pests and diseases.

Rootstocks breeding objectives

- To develop rootstocks capable of surviving under wide range of environmental condition.
- To develop rootstocks that are easy to root.
- To develop rootstocks having resistance against pest and disease.

14.9 Breeding Methods

14.9.1 Introduction

In apple, a lot of emphasis had been laid on the introduction and selection during the last few decades. Accordingly, several cultivars have been introduced through NBPGR, New Delhi, e.g., Red Spur Delicious, Golden Spur Delicious, Royal Red, Vared, Michael etc.

14.9.2 Clonal Selection

Red Delicious was evaluated as a chance seedling in Iowa, USA. 'Kodaikanal Beauty' was developed by clonal selection from 'Parlin's Beauty'.

14.9.3 Hybridization

Apple breeding work has been going on at Regional Fruit Research Station, Mashobra, Himachal Pradesh, Fruit Research station, Shalimar, Jammu and Kashmir and Horticulture Experiment and Training Centre, Chaubattia, Uttar Pradesh resulting in release of a number of cultivars, e.g., Lal Ambri (Red Delicious × Ambri), Suneheri (Ambri × Golden Delicious), Ambred (Red

Delicious × Ambri 57), Chaubattia Princess (Red Delicious × Early Shanbury) etc.

14.9.4 Mutation Breeding

Bud sport has resulted in development of many new cultivars in apple. 'Red Elstar' is a natural mutant of the parent cultivar 'Elstar'.

14.10 Cultivars

Season	Apple Cultivars
Early-season	Irish Peach, Benoni, Tydeman's Early Worcestor (P), Michael, Mollies Delicious, Schlomit, Starkrimson, Anna, Early Shanburry, Fenny, Chaubattia Princess.
Mid-season	American Mother, Razakwar, Jonathan (P) Cox' Orange Pippin (P), Queen's Apple, Rome Beauty, Scarlet Siberian, Starking Delicious, Red Delicious, Rich-a-red, Vance Delicious, Top Red, Lord Lambourne, Red Chief, Oregon Spur, Red Spur, Red Gold (P), Silver Spur, Scarlet Gala, Well Spur, Super Chief, Scarlet Spur, McIntosh, Cortland, Golden Delicious.
Late-season	King Pippin, American Apirouge, Kerry Pippin, Lal Ambri, Sunheri Chamure, Ambri, Baldwin, Yellow Newton, Golden Delicious (P), Yellow Newton (P), Winter Banana, Granny Smith (P), Red Fuji, Coe Fuji, Rymer, Buckingham.

Spur types and colour sports: Red Chief, Oregon Spur-II, Silver Spur, Well Spur, Red Spur, Super Chief, Starkrimson, Hardi Spur., Schelet Spur, Ace Spur.

Standard colour mutants: Vance Delicious, Top Red, Skyline Supreme, Hardiman, Bright-N-Early.

Standard cultivars: Starking Delicious, Red Delicious, Rich-a-red.

Low chilling cultivars: Michael, Schlomit, Anna, Vered, Tamma, Tropical Beauty and Parlin's Beauty.

Pollinizing cultivars: Golden Delicious, Red Gold, Tydeman's Early Worcester, Summer Queen, Golden Spur, Granny Smith, Winter Banana, Mclntosh, Scarlet Gala and flowering crabs like Manchurian, Snow Drift and Malus floribunda

Scab resistant cultivars: Prima, Priscilla, Sir Prize, Florina, Flrdous, Shireen, Macfree, Coop-12, Coop-13, Liberty and Freedom.

Hybrid cultivars: Lal Ambri (Red Delicious × Ambri), Sunheri (Ambri × Golden Delicious) Chaubattia Princess (Red Delicious × Early Shanbury) and Chaubattia Anupam (Early Shanburry × Red Delicious), Ambred (Red Delicious × Ambri), Ambrich (Richared × Ambri), Amroyal (Starking Delicious × Ambri).

Source: Hand Book of Horticulture, ed. by Dr. K.L.Chadha

Review Questions

Part A

Answer the following questions true or false.

1. The centre of origin of apple is from Asia Minor to Western Himalaya. True/False
2. Ancestor of apple is *Malus sylvestris.* True/False
3. The apple trees do not require chilling hours to break the rest period. True/False
4. In self-incompatible cultivar there is need of emasculation. True/False
5. Red Spur Delicious is an introduction to India. True/False

Part B

Write a brief note on each of the following.

1. Breeding objectives of apple.
2. Breeding methods in apple.
3. Emasculation and crossing technique in apple.

15

Breeding of Pear

15.1 Introduction

Pear is one of the important fruit crops that belong to the family Rosaceae. In the world pear is cultivated in China, USA, Argentina, Italy, Turkey, Spain, India, South Africa, Japan, Belgium etc. It is grown under a wide variety of climatic regimes ranging from cold dry temperate hilly conditions to warm humid sub-tropical conditions on the plains of northern India. In India pear is widely cultivated in Punjab, Himachal Pradesh, Jammu and Kashmir, Uttar Pradesh, Uttarakhand etc.

15.2 Origin

The primary centre of origin is China and Asia Minor to the Middle East in the Caucasus mountains. Secondary centre is in Central Asia. The ancestors of *P. communis* are considered to be *P. communis* var. *pyraster*, *P. communis* var. *caucasica* and probably *P. nivalis* (it is known as snow pear).

15.3 Taxonomy and Classification

The taxonomic position of pear as per Hutchinson system is:

Phylum : Angiospermae

Sub-phylum : Dicotyledones

Division : Lignosae

Order : Rosales

Family : Rosaceae

Genus : *Pyrus*

Species : *communis*

Binomial name : *Pyrus communis* L.

Pear has a chromosome number of 2n=34.

15.4 Morphology

***15.4.1 Habit*:** Deciduous, small to medium sized tree, 10 m tall (normally 3-5 m in cultivation)

***15.4.2 Root*:** Tap root system.

***15.4.3 Stem*:** Conical erect trunk, branches small, reddish-brown and narrow angled, grey-brown bark with shallow furrows and flat-topped scaly ridges.

***15.4.4 Leaf*:** Petiolate, stipulate, simple, alternate, elliptic/ovate, finely serrated margin, obtuse apex, 2.5-10 cm long and 3-5 cm wide, shiny green above and paler and dull below, glabrous.

***15.4.5 Inflorescence*:** Corymbose, 5-7.5 cm wide, with 5-7 flowers.

***15.4.6 Flower*:** Snowy white, 2.5-3.5 cm wide, borne from terminal, mixed buds of short spurs, appearing before or with the leaves.

15.4.7 Calyx: Sepals-5

15.4.8 Corolla: Petals-5

***15.4.9 Androecium*:** 20-30 stamens with usually red anthers.

15.4.10 Gynoecium: Ovary inferior, pentacarpellary, 5-locular, 2 ovules per locule.

***15.4.11 Fruit*:** Pyriform pome with persistent or delicious calyx, 4-12 cm long, greenish coloured.

15.4.12 Seed: Blackish, 8.4 × 4.8 mm, each with a thin layer of endosperm.

15.5 Floral Biology

The lowest mean daily temperature for the start of flowering in pears is 9 ^{0}C. The length of flowering period is dependent upon the mean daily temperature and atmospheric humidity. The higher the mean temperature and lower the humidity, the shorter the flowering period. Pear trees begin flowering at the age of 5-8 years mostly on spur and terminal shoots. Pear bears fruits on 2-year-old shoot. Floral bud differentiation starts in July while bloom occurs next year in spring.

15.6 Pollination

Homomorphic, gametophytic, monofactorial, multiallelic, SI system is present in pear. *Pyrus communis* requires pollinizers to get economic crop yield. Pollinizer is cross-compatible with commercial cultivars and SI allelic constitution in both cultivars has to be different in at least one of the alleles.

Cross-pollination is considered essential for the production of good crop. Beurre Hardy, Comice, Patharnakh, Flemish Beauty are self-fruitful; whereas, Le Conte, MAgness, Winter Nelis are self-unfruitful. And majority of other commercial cultivars are partially self-fruitful. When only one commercial cultivar is under cultivation, the placement of pollinizers becomes crucial. In such situation, planting of every fourth tree in every fourth row as pollinizer is adequate. General characters of good pollinizers: they have sufficient overlapping of blooming period, should bear fruit of commercial value, and should be fire blight resistant. Placing of two colonies of honey bees per ha is beneficial for effective pollination. China is a good pollinizer cultivar of pear. Flowering time of early cultivars is 1st week of March, for mid-season cultivar 3rd week of March and for late season cultivar 4th week of March to 1st week of April.

15.7 Emasculation and Crossing Techniques

One to three flowers/cluster at the 'balloon stage' of development are selected and all anthers are removed. At least two flowers must be emasculated and pollinated in order to have at least one seed. Emasculation is done using finger nails, scissors, scalpels or tweezers. A cut is made below sepal and flower is pulled off leaving only the pistil. Branches with full bloom are collected at tight cluster stage of bud development with their cut ends in water, 2-3 days prior to pollination. Anther harvesting is done by rubbing the flowers over a wire mesh screen. Anthers are dried for 24 hours and the pollen can be stored for over a year with refrigeration. Pollen transfer is made through the use of camel hair brush. Bees do not visit flowers without corolla and so emasculated and pollinated flower does not require bagging. Pears require stratification.

15.8 Breeding Objectives

- To develop dwarf scion.
- To develop dwarf rootstocks tolerant to wet and saline soils.
- To develop rootstock resistant to diseases like Ganoderma and root rot, and pests.
- Solution to problems of low bud differentiation, alternate and shy-bearing of Baghugosha and Le Conte.
- Selection of superior clones of Patharnakh and Baghugosha.

15.9 Breeding Methods

15.9.1 Introduction

Important and popular cultivars such as Bartlett, Anjou, Kieffer are only introduction from Europe and are well acclimatized to the Northern and Southern Indian hills.

15.9.2 Mutation Breeding

Spontaneous mutation (bud sports) has given red coloured cultivars 'Starkimson' from Clapp's Favourite and Red Anjou from Anjou. Spotting of coloured fruit mutants is easier than others in pears.

15.10 Cultivars

Cultivars grown in Jammu and Kashmir:

William, Kashmir Nakh, Vicar of Wakefield, Beurre D. Amanlis, Goshbagu, Beurre Hardy.

Cultivars grown in Himachal Pradesh:

China, Bartlett, Max Red Bartlett, Flemish Beauty, Devoe, Mannings's Elizabeth, Winter Nellis.

Cultivars grown in Uttar Pradesh:

Thumb Pear, Doyene du Comice, Victoria, William's Bartlett, Beurre Hardy, Flemish Beauty.

As per period of maturity the cultivars are:

Sl.	Group	Period of Maturity	Cultivars
1	Extreme early	1-15th June	Early China
2	Early	16-30th June	Basuiodse Favourite
3	Mid early	1-15th July	Red Bartlett, Max Red Bartlett, Starkrimson, Kashmir Boggugosha
4	Mid-season	16-31st July	Bartlett, Flemish Beauty, Victoria, Starking Delicious, Beurre Bosc
5	Late mid-season	1-15th August	Smart Pear, Thumb Pear, Monarch
6	Late season	16-31st August	Common Pear, Country Pear, Beurre Hardy, Kashmir Pear
7	Very late	1-15th September	Awal Number, New Pear, Weldon
8	Extreme late	16th September onward	Conference

Review Questions

Part A

Answer the following questions true or false.

1. Pear is grown under a wide variety of climatic regimes. True/False
2. Pear has a chromosome number of 2n=36. True/False
3. The lowest mean daily temperature for the start of flowering in pears is 9^0c. True/False
4. Emasculation is done using finger nails in pears. True/False
5. Bartlett is an introduction from the USA. True/False

Part B

Write a brief note on each of the following.

1. Floral biology and pollination in pear.
2. Breeding methods in pear.
3. Emasculation and crossing technique in pear.
4. Breeding objectives of pear.

16

Breeding of Peach

16.1 Introduction

Peach is one of the important fruits grown in warm temperate zones of the world. Fruits are rich in protein, sugar, minerals and vitamins. Peaches are highly valued as table fruit for their attractive colour and palatability. Canned peaches, dried frozen preserves, jams, nectar are principal processed products of peach. It is grown mainly in Jammu and Kashmir, Himachal Pradesh, Punjab, Uttarakhand, Nilgiri hills, Jharkhand and North-Eastern States.

16.2 Origin and Distribution

China is the centre of origin of peach. It was domesticated there 4000 - 5000 years ago. The main routes of peach movement from China to the West were sea through India and Mid-East as well as silk route through Persia.

16.3 Taxonomy and Classification

The taxonomic position of peach as per Hutchinson is:

Phylum : Angiospermae

Sub-phylum : Dicotyledones

Division : Lignosae

Order : Rosales

Family : Rosaceae

Genus : *Prunus*

Species : *persica*

Binomial name: *Prunus persica* (L.) Batsch.

Peach is a diploid (2n = 2x = 16).

P. mira, P. dividiana, P. fergamensis are the wild species of peach. *P. behmi* is the natural hybrid between almond and peach. Nectarines are botanical variety of peach and classified as *P. persica* var. *nucipersica.* Nectarines are group of peaches in which the fruit surface does not have fuzzy skin and are usually smaller in size and have greater aroma. This has developed from peach as bud sports in which single recessive gene responsible for fuzziness is absent.

16.4 Morphology

16.4.1 Habit: Deciduous, small tree with a spreading canopy, usually 6 to 10 feet, short lived generally living only 15 to 20 years as productive life.

16.4.2 Root: Tap root system.

16.4.3 Stem: Aerial, erect, branched, cylindrical, solid, woody.

16.4.4 Leaf: Sub-sessile, exstipulate, many pair of glands at the base of each leaf, alternate, simple, oblong/lanceolate, 10-12 cm long, acute apex, finely serrated margin, folded slightly along the mid-rib, glabrous.

16.4.5 Inflorescence: Solitary flower or in cymose at the tip of shoot.

16.4.6 Flower: Sub-sessile or pedicellate, complete, actinomorphic, perigynous, thalamus cup shaped, pentamerous, light pink to carmine to purplish, appearing before the leaf emergence.

16.4.7 Calyx: Sepals-5, gamosepalous, calyx tube campanulate, caducous, green, woody.

***16.4.8* Corolla:** Petals-5, polypetalous, pink, imbricate.

16.4.9 Androecium: Stamen numerous (10-60), polyandrous, inserted at the ridge of calyx tube, anther dehiscence longitudinal.

16.4.10 Gynoecium: Ovary half-inferior, monocarpellary, unilocular, placentation marginal, single ovule, simple style, stigma capitate/clavate.

16.4.11 Fruit: Single seeded drupe, fruit widely variable in size and shape, velvety skin with yellow dark red colour, flesh white or yellow, stone very hard and deeply pitted or furrowed.

16.4.12 Seed: Non-endospermic, seed may be free, semi-clinging or clinging type with flesh.

16.5 Floral Biology

Peach is precocious tree commencing bearing 2nd -3rd year after planting. Fruit buds are borne mainly laterally on one year old wood from terminal and axillary vegetative flush/buds. Generally, 3 buds arise at a node, the side buds being floral and central is vegetative. Flowering time is March in India but some low chilling cultivars gives flower during late January. The anthesis occurs between 5 am and 11 am. The stigma remains receptive after 2-4 days of anthesis. Flowering period in peaches is determined by the completion of chilling requirement and prevailing temperature during early spring.

16.6 Pollination

Self-pollination (autogamy) has been reported in all the peach cultivars studied. Sometimes fruit set was higher in bag than open-pollinated flowers. Practically all peach cultivars are self-fertile except J.H. Hale, June Elberta, Halberta, Candoka, Chinese Cling and Giant.

16.7 Emasculation and Crossing Techniques

Pollen is collected at the balloon stage when the flowers are not quite open. Anthers are collected by rubbing flowers over a wire mesh screen. At room temperature pollen can be stored in glass vials for a season. Branches are emasculated from top to bottom to avoid any accidental wind pollination. From bisexual flowers stamens can be easily removed. Emasculated flower is not covered as pollinators are not attracted to petal less flowers. Branches bearing pollinated flowers are however covered with paper bag or cheese cloth to avoid any chance pollination. Seeds are collected from ripe fruits and are stratified before drying. Seeds kept in moist perlite in plastic bag are stratified and after that temperature required for germination is provided. Germinated seeds with 0.5 - 1.0 cm long radicle can be transferred to field or transferred to green house for further growth and then to field. Use of embryo culture can be used to germinate seeds obtained from early maturing cultivar in sub-tropical areas. In case of early maturing cultivars, generally the flesh matureds before the embryo is fully developed.

16.8 Breeding Objectives

For scion

- To develop cultivar with low chilling requirement.
- Extension of season of maturity and early maturing cultivars harvest before rain in sub-tropical areas.
- Tolerance to high summer temperature.
- Maturity between 60-70 days after full bloom.

- Alternate colour, firm flesh, freedom of stone from pulp and non-browning of flesh.

For rootstock

- Resistant to root knot nematodes, iron chlorosis and water logging.
- Control of tree vigour for HDP.
- Ease of propagation and stock-scion interaction with higher graft compatibility.
- Wide range of soil and climatic adaptation.

16.9 Breeding Methods

Systematic breeding work on improvement of peaches was initially started in China. In India planned breeding programme was initiated in 1957 at Saharanpur. Some work was initiated at Pantnagar but no cultivar was found suitable for commercial cultivation. Later on, rigorous breeding work was carried out at Solan and Ludhiana.

16.9.1 Introduction

A large number of low chilling cultivars, e.g., Flordasun, Sun Red, Sun Gold and some other selections were introduced successfully. Flordared and Flordabelle were introduced at PAU, Ludhiana during late sixties from Florida and California. Of these introductions, Flordasun, Sun Red, Flordared and 16-33 (later named as Shan-e-Punjab) became very popular in India. TA170, popularly known as 'Pratap' and an introduction from USA, has been identified as early (7 days earlier than Flordasun). Another two introductions from Florida, Flordaprince and Earligrande have been recommended for the plains of Punjab and adjoining areas.

16.9.2 Clonal Selection

'Sharbati' is a chance seedling selected at Saharanpur.

16.9.3 Hybridization

Redhar is a cross between Halehaven and Kalhaven bred at USA. Nemagcrad, a hybrid between *P. Persica* and *P. davididasa* is widely used root-knot nematode resistant rootstock, which is immune to *Meloidogyne incognita*. Planned hybridization work on peach was started in 1957 at Saharanpur. Peach Saharanpur Prabhat (Sharbati × Flordasun) was released.

16.9.4 Mutation Breeding

Two cultivars have been released through γ-irradiation, namely, 'Magnof 135' and 'Plovdiv 6'. Bud sports early ripening mutant from Redhaven have been identified, namely, 'Garnet Beauty' and 'Stark Eali Glo'.

16.10 Cultivars

Sharbati: Clingstone, medium fruit, greenish skin, white flesh.

Pratap: Freestone, fruit medium, red blush skin, semi-vigorous tree, mid-season.

Prabhat: Clingstone, fruit medium, skin yellow with green tinge, semi-vigorous tree, early.

Shan-e-Punjab: Semi-clingstone, large fruit, skin yellow with red spot, vigorous tree, late.

Review Questions

Part A

Answer the following questions true or false.

1. China is the centre of origin of peach. True/False
2. Peach is a triploid. True/False
3. Sharbati is a chance seedling selected at Saharanpur. True/False
4. Self-pollination has been reported in all the cultivars studied. True/False
5. Pratap cultivar is clingstone. True/False

Part B

Write a brief note on each of the following.

1. Breeding methods of peaches.
2. Breeding objectives in peaches.
3. Emasculation and crossing techniques in peaches.

17

Breeding of Almond

17.1 Introduction

Almond or *badam* is a high energy nut. Almond kernels contain lipids (52.2%), protein (20%), carbohydrate (20.4%) and water-soluble sugars (4.4%). Irrigation affects almond yield and quality. Almond quality is primarily genotype dependent. Maturity at harvest can also influence kernel quality.

17.2 Origin

The origin of almond is Central Asian mountainous region. Almond appears to be a hybrid between *Prunus fenzliana* and *Prunus bucharica.*

17.3 Taxonomy and Classification

The taxonomic position of almond as per Hutchinson is:

Phylum : Angiospermae

Sub-phylum : Dicotyledones

Division : Lignosae

Order : Rosales

Family : Rosaceae

Genus : *Prunus*

Species : *amygdalus*

Binomial name: *Prunus amygdalus* Batsch
Syn. *Prunus dulcis* (Mill.) D.A. Webb

It is a diploid (2n=2x=16).

17.4 Pollination

There is presence of heterostyly (SI). It is a cross-pollinated crop. Pollination is carried out through honey bees.

17.5 Breeding Objectives

- Late blooming behaviour (to escape spring frost damage to blossoms)
- Cross-compatibility or self-fruitfulness
- Heavy bearing
- Superior nut and kernel quality

17.6 Breeding Methods

17.6.1 Selection

There is a lot of natural variability available in the seedling plantations of almond. Intensive surveys and evaluation led to the release of cultivars, namely, Makhdoom, Parbat, Waris and Shalimar in Jammu and Kashmir. They escape spring frost, are cross-compatible and have good bearing potential under rainfed condition.

17.6.2 Hybridization

Hybridization programme was started in PAU, Ludhiana in 1970 to evolve cultivars for subtropical plains. A large number of almond cultivars were crossed with Sloh (Peach × Almond hybrid). Hybrid 15 (Thin Shelled × Sloh), Hybrid A258 (Pethick's Wonder × Sloh) and H-98 (a cross of Bruce × Sloh) were released.

17.6.3 Mutation Breeding

A late flowering, self-compatible almond named 'Supernova' was derived from 'Fascionello' through γ-irradiation following selection in MV_2.

Review Questions

Part A

Answer the following questions true or false.

1. The origin of almond is Central Asian mountainous region. True/False
2. Almond is a self-pollinated crop. True/False
3. Almond is a triploid. True/False
4. There is no heterostyly (SI) in almond. True/False
5. Pollination is carried out through honey bees in almond. True/False

Part B

Write a brief note on each of the following.

1. Breeding methods in almond.
2. Breeding objectives in almond.

18

Breeding of Walnut

18.1 Introduction

Among nut crops, 'Walnut' or '*Akhrot*' is the most important crop. It has immense commercial importance even in the areas situated in remotes without having good market and transport infrastructure. Because of long storability, growers have opportunity to hold the produce at room temperature for two to three months easily and sell out at intended time. Walnut is rich in protein, fats and vitamins, especially vitamin B_6. The walnut kernel contains about 90% PUFA of which 26% is the Omega-3 fatty acid, alpha-linolenic acid. But the problem with this type of kernel is that higher the percentage of PUFA in it, the more is the chances of getting the oil oxidized and thereby reducing the shelf life of the walnut kernel.

18.2 Origin

Walnut is native to South-East Asia.

18.3 Taxonomy and Classification

The taxonomic position of walnut as per Hutchinson is:

Phylum : Angiospermae

Sub-phylum : Dicotyledones

Division : Lignosae

Order : Juglandales

Family : Juglandaceae

Genus : *Juglans*

Species : *regia*

Binomial name: *Juglans regia* L.

Walnut is diploid (2n=2x=32).

18.4 Pollination

Plants are monoecious. It is self-fertile and cross-compatible. There are cultivars that are protogynous in nature. Also, some cultivars are protandrous. Cross-pollination is through wind-borne pollen.

18.5 Breeding Objectives

- Development of cultivars with more number of female flowers.
- To develop cultivars with less flower and fruit drop.
- To develop cultivars with regular bearing habit.
- To develop high quality medium shelled nuts.
- To develop nuts with thin shell and high kernel recovery percentage and extra light-coloured kernels for shelling purpose.

18.6 Breeding Methods

Selection

Most of the walnut trees in India are of seedling origin. A few promising seedlings have been selected in Jammu & Kashmir, Himachal Pradesh and Uttar Pradesh. As a result, a selection 'Karan' (also known as 'Bulbul dun') having papery shell with high shelling percentage (more than 60 % kernel recovery) was identified in Kashmir.

From hills of Himachal Pradesh and Uttar Pradesh, superior genotypes, namely, Govind, Roopa and Ratna were selected. Chakrata No. 14 was found the best with regard to thin shell, good filling and attractive light amber colour of kernel in Chakrata hills and Jaunsar Bhabhar areas of Uttar Pradesh. The main constraints in breeding are long selection period and giant size of tree.

Central Institute of Temperate Horticulture (CITH), Srinagar has identified many high yielding clones from seedling population and are supplied as grafts to the farming community.

Review Questions

Part A

Answer the following questions true or false.

1. Walnut is native to South-East Asia. True/False
2. Walnut is diploid (2n=2x=34). True/False

3. Cross-pollination in walnut is through wind-borne pollen. True/False
4. Development of cultivars with more number of female flowers is not desirable in walnut. True/False
5. Plants of walnut are monoecious. True/False

Part B

Write a brief note on each of the following.

1. Breeding objectives in walnut.
2. Selection in walnut.

SECTION III

BREEDING OF PLANTATION CROPS

19

Breeding of Coconut

19.1 Introduction

Coconut is valued most for its nut. The other components are coconut water, kernel, shell and husk. Coconut water is one of the most refreshing drinks. It is also known for its medicinal value. It increases blood circulation in kidneys, causes profuse diuresis and eliminates mineral poisoning. It is a good substitute for saline glucose under gastroenteritis conditions. The coconut has been regarded as 'Kalpavriksha' because all the parts of the palm is useful to mankind in one way or other. The dried kernels commonly called as copra is the richest source of vegetable fat (60-67 % oil). Coconut is major source of edible oil. India is the largest producer of coir and its product. The major coconut growing areas of the world are found in Asia, Oceania, West Indies, Central South America, East and West Africa. In India, coconut is commercially grown in Kerala, Karnataka, Tamil Nadu, Andhra Pradesh, Gujarat, Odisha and West Bengal.

19.2 Origin and Distribution

Coconut is originated and domesticated in Malaysia. The precise centre of origin of coconut is still a disputed issue with two distinct zones of diversity, i.e., South Pacific Islands of Polynesia and Melanesia and South East Asia and South America. Its pan tropical distribution is limited to 23° N and 23° S of equator.

19.3 Taxonomy and Classification

The taxonomic position of coconut as per Hutchinson is:

Phylum	:	Angiospermae
Sub-phylum	:	Monocotyledones
Division	:	Corolliferae
Order	:	Arecales (syn. Palmales)

Family : Arecaceae (syn. Palmae)

Genus : *Cocos*

Species : *nucifera*

Binomial name : *Cocos nucifera* L.

Cocos is a monotypic genus with a single species *nucifera*. Hence, there is a rather restricted genetic base with no known wild or domesticated relatives. This limits the possibility for tapping gene pools from related sources.

It is a diploid with 2n= 2x= 32.

19.4 Morphology

19.4.1 Habit: Tall, unbranched with a terminal crown of leaves, height 20-30 m.

19.4.2 Root: Radicle is temporary and penetrates soil vertically. Subsequent ones spread horizontally. Adventitious roots are produced from bole, basal 60 cm swollen portion of stem. An adult palm has 2000-3000 living main roots, confined to top 1.5 m from ground level. Main roots branch occasionally and are cord like.

19.4.3 Stem: Coconut palm has only one growing point. No visible trunk is formed until apical meristem has attained its full diameter. Trunk is formed after 3 years, i.e., after production of 12-18 leaves. Once apical meristem has attained its maximum size, there will not be any further increase in diameter. If a stem is damaged, wounds cannot be repaired as there is no cambium.

Trunk is columnar and erect, light, greenish brown and 20-40 cm in diameter. Stem has nodes bearing a leaf or scar if that leaf has fallen already. Internodes are very short. About 12-14 leaves are shed every year and thus age of palm can be ascertained by counting number of scars and dividing by 12 (i.e., total number of leaves/year) and adding three years for trunk formation.

19.4.4 Leaf: Leaves are borne in a terminal radiating crown consisting of 25-35 opened leaves and a central bud with leaves in various stages of development. A fibrous leaf sheath protects young leaf until it matures. Life span of a leaf is 2½ to 3 years. A healthy adult palm has 8-10 leaves with nut harvested, 10-14 leaves with nuts of varying maturity, 10-12 opened leaves with spadices in different stages of growth and central bud.

Leaves are arranged in 5 spirals with a phyllotaxy nearly 2/5 with an angle of about 140° between successive leaves. Developing leaves are perpendicular but are gradually pushed aside as new leaves develop. Leaves are large, paripinnate, 4.5-6 m long, weighs about 10-15 kg. Petiole is stout with clasping sheath at base, attached to trunk. Number of pinnae is about 200-250 along upper 3 quarters of petiole, with apex acute and sessile.

***19.4.5 Inflorescence*:** Palm is a monoecious with numerous male and female flowers in each spadix, borne singly in axil of each leaf of a bearing palm. Flowering commences at 6-12 years of age. In regular bearers, number of leaves and number of spadices are same, i.e., 12/year.

***19.4.6 Flower*:** Inflorescence consists of a central axis with about 40 lateral branches called spikelets. Each spikelet bears about 200-300 male flowers at top, opening from tip downwards, and one or more female flowers at base. Male flowers borne singly or 2 or 3 in a bunch, sessile, pale yellow. Female flowers are globose, 2-3 cm in diameter.

***19.4.7 Perianth*:** 6 tepals (perianth leaves) arranged in two whorls of three each, the outer being smaller, imbricate, persistent in female flowers.

***19.4.8 Androecium*:** 6 stamens, arranged in two whorls, filaments free, anther versatile dehiscing by vertical split.

***19.4.9 Gynoecium*:** Ovary superior, tricarpellary, uni-/bi-/tri-locular, syn-carpous, rarely apocarpous, axile placentation, style short, trifid stigma.

***19.4.10 Fruit*:** It is a fibrous drupe, weighing 1.2-2.0 kg and not of commercial value, consists of seed and endocarp. Matured fruit is ovoid in shape, three angled, varying size, taking 12 months to mature, fruit consists of an exocarp (outer skin) which is tough, fibrous mesocarp 4-8 cm in thickness and gives coir, and endocarp which is ovoid shell enclosing a thick albuminous endosperm or meat and a single seed. In between shell and endosperm, there is a brown testa adhering firmly to endosperm.

19.5 Floral Biology and Pollination

In coconut palm, especially in tall cultivars, there is a distinct gap between male and female phases. Female flowers do not become receptive until all male flowers in same spadix have shed their pollens. Cross-pollination is customary. Slight chances of self-pollination also exist from succeeding inflorescences during summer. In dwarf cultivars and hybrids, interval between two phases is less and narrow, thereby increasing chances of self-pollination. Maturation of inflorescences is a progressive process, pollen discharge and anthesis continues for 18-20 days. Male flowers after opening shed pollen for about 24 hours and then drop off. Female flowers commence to open by 21st day, but during summer they may open even at 18th day itself and continue receptivity for 4-7 days thereafter. When stigma is receptive, nectar is secreted from female flowers. The period of stigma receptivity varies in different places with normal range of 1-3 days.

Pollination in coconut is effected through wind and insects. Among insects, bees are major pollinating agents.

19.6 Emasculation and Crossing Techniques

The male flowers are removed before opening of female flowers. There are hundreds of male flowers and few female flowers in an inflorescence. All the male flowers are removed to avoid any self-pollination by cutting the spikes about 4-5cm away from the last female flower. The remaining male flowers are removed by hand.

The male flowers on the top and middle portion of spike produce more fertile pollens. The mature stamens with plenty of pollens will have a bluish green tinge at the tip. The male flowers with mature stamens are collected and processed as follows:

- Place the male flowers between folds of thick papers and gently crush them with a rolling pin.
- Keep the crushed male flower in an oven at 40^0C for 24 hours.
- Using a 0.2 mm mesh sieve separate the pollens from the debris.
- Test germination of pollens on a sugar, agar and gelatin medium (8:2:2). Only pollens with at least 50% germination should be used for pollination.
- Collect pollens in glass vials and stir in desiccators over fused calcium chloride ($CaCl_2$). Pollens can be used for 10-15 days.
- If longer storage is required, seal the glass vials and store them in deep freezer (-20^0C). The pollens can be used for 3-5 months.
- Dilute the pollens with neutral talc powder in 1: 9 proportion before use. If the pollens stored in deep freezer is to be used, first allow them to come to room temperature before diluting.

The female flowers should be bagged for the entire period of female phase for pollination with the desired pollens. The receptivity of stigma is indicated by the presence of nectar at three places below the stigma and also by splitting of the stigma into three lobes. When the stigma turns brown or black, it is no more receptive. Pollen applicator consisting of a polythene bottle, a rubber tube and a bamboo pole can be used for pollination. The pollen talc mixture is taken in the bottle which is squeezed by pressing the rubber tubing. Pollination can also be repeated on the same flower for ensuring crossing. The setting percentage of manually crossed flower is found to be higher than the natural setting.

19.7 Breeding Objectives

- To evolve cultivars with higher net production.
- To evolve cultivars with higher copra and oil content.
- To evolve cultivars possessing drought tolerance.
- To evolve cultivars resistant to important diseases, especially to root wilt and pests.
- To evolve cultivars exclusively for use as tender nuts.

19.8 Breeding Methods

An organized breeding of coconut was started for the first time in the world in 1916 at Coconut Research Stations at Kasaragod and Nileshwar (Pilicode) of former Madras Presidency (now in Kerala state).

Important breeding methods used in improvement of coconut are as follows:

19.8.1 Introduction

The different cultivars of coconut from Ceylon, Indo-China, New Guinea, Java, Indonesia, Thailand, Philippines, Fiji, Laccadives etc. were introduced in India by Madras Agricultural Department. The Cochin Department of Agriculture had also introduced cultivars from Malaysia, Sea Island and Philippines and recently cultivars from Solomon Islands, Borneo, Seychelles, Panama, East Africa, West Indies etc.

Important cultivars introduced from different countries are Java, New Guinea, Cochinchines, Philippines, Laccadive Ordinary, Laccadive Micro, West Coast Tall etc.

19.8.2 Selection

Mass selection can be used as effective method of improvement of coconut. In this method, superior mother palms are selected. The success of this method depends on the ability of the breeders and heritability of characters under consideration.

Examples: Pratap, Lakshadweep Ordinary.

19.8.3 Hybridization

The first coconut hybrid in the world was produced in 1934 by Dr. J. S. Patel with West Coast Tall as female and Chowghat Green Dwarf as male parents. The following cross combinations have been attempted in coconut:

Dwarf × Dwarf

Tall × Dwarf

Dwarf × Tall

Tall × Tall

19.8.3.1 Dwarf × Dwarf (D × D)

The crosses between D × D have not given satisfactory result with respect to yield potential traits.

19.8.3.2. Tall × Dwarf (T × D)

The hybrids possess desirable traits such as precocity on bearing, higher productivity than parents. The important dwarf cultivars used as male parents are Dwarf Orange, Dwarf Green, Gangabondam and Malayan Dwarf Yellow. All the hybrids of Tall × Dwarf Green, Tall × Dwarf Orange and Tall × Gangabondam showed heterosis in the weight of nut, kernel content and nut water. Among the male parents, Dwarf Orange and Gangabondam proved to be best for production of economic hybrids with West Coast Tall. The superior performance of Gangabondam as male parent has been reported from Kasaragod, Mahwa and Coimbatore. In comparative trials the superior performance of Lakshadweep Ordinary × Chowghat Orange Dwarf and Lakshadweep Ordinary × Gangabondam at Kasaragod has been reported. Chowghat Orange Dwarf and Gangabondam are the desirable male parents and West Coast Tall and Lakshadweep Ordinary are desirable female parents.

Examples: Chandra Laksha, Laksha Ganga, VHC 1, VHC 2, Kera Sankara, Anandaganga, Kera Ganga, Kera Sree.

19.8.3.3 Dwarf × Tall (D × T)

These hybrids have been found to have higher production potential than T × D hybrids. The Chowghat Orange Dwarf × West Coast Tall hybrid was found to have better nutrient utilization ability and responded well to lower levels of applied fertilizers.

Example: Chandra Shankara

19.8.3.4 *Tall × Tall (T × T)*

The poor yield in tall combinations may be due to the poor combining ability of the parents. Thus, it needs to be emphasized to select promising tall parents based on combining ability of traits.

19.9 Cultivars

19.9.1 Important Coconut Cultivars

Tall Group: West Coast Tall (WCT), East Coast Tall (ECT), Tiptur Tall (TPT), Benaulim Tall (BENT) or Pratap, Lakshadweep Micro (LMT), Lakshadweep Ordinary (LOT), Andaman Ordinary (ADOT), Philippines Ordinary (PHOT).

Dwarf Group: Chowghat Orange Dwarf (COD), Chowghat Green Dwarf (CGD), Gangabondam Dwarf (GBD), Malayan Yellow Dwarf (MYD).

19.9.2 Hybrids

Chandrasankara (COD × WCT), Kerasankara (WCT × COD), Chandralaksha (LOT × COD), Lakshaganga (LOT × GBD), Keraganga (WCT × GBD), Anandaganga (ADOT × GBD), Kerasree (WCT × MYD), Kerasowbhagya (WCT × SSAT), VHC 1 (ECT × DG), VHC 2 (ECT × MYD), Godavariganga (ECT × GBD)

19.9.3 Cultivars released by CPCRI and SAUs

Chandrakalpa, Kerachandra, Kalpa Pratibha, Kalpa Mitra, Kalpa Dhenu, Chowhat Orange Dwarf, Gautamiganga, Kalparaksha, Kalpasree, Pratap, VPM 3, ALR 1, Kamrupa, Kera Sagara, Kera Bastar, Kera Keralam, Kalyani Coconut, Kalpatharu.

Review Questions

Part A

Answer the following questions true or false.

1. Coconut is originated and domesticated in Malaysia. True/False
2. *Cocos* is not a monotypic genus. True/False
3. The crosses between D × D have not given satisfactory result. True/False
4. The poor yield in T × T crosses may be due to the poor combining ability of the parents. True/False
5. Coconut is a triploid. True/False

Part B

Answer the following questions.

1. Cite two hybrids of coconut.
2. D × T hybrids have higher/lower production potential than T × D hybrids.
3. Chandrashankara is a —.
4. Coconut belongs to family —.
5. What is the somatic chromosome number of coconut?

Part C

Write a brief note on each of the following.

1. Breeding methods in coconut.
2. Emasculation and crossing techniques in coconut.
3. Breeding objectives of coconut.
4. Floral biology and pollination of coconut.
5. Cultivars of coconut.

20

Breeding of Arecanut

20.1 Introduction

Arecanut (or betel palm) is chewed as narcotic either alone or more commonly along with betel leaf. Its nut has medicinal and pharmaceutical properties. About $1/10^{th}$ of world population has a habit of betel chewing. Nut contains polyphenol (11-18 %) which includes flavonols like catechin, epicatechin, leucocyanidin, fat (9-15%), polysaccharides (17-25%), fibre and protein (6-7.5%) and some minerals like calcium, phosphorous and iron.

20.2 Origin

Many reports suggested that this crop has been originated from the Philippines, Malay, Indonesia, Cochin, China and Malaya peninsula where this has been widely used for religious purpose.

20.3 Taxonomy and Classification

The taxonomic position of arecanut as per Hutchinson is:

Phylum : Angiospermae

Sub-phylum : Monocotyledones

Division : Corolliferae

Order : Arecales (syn. Palmales)

Family : Arecaceae (syn. Palmae)

Genus : *Areca*

Species : *catechu*

Binomial name : *Areca catechu* L.

Areca is given the status of a monospecific genus. The genus expanded rapidly from its mono-specific status and at present contains about 76 species. *Areca catechu* is the only cultivated species. Its chromosome number is 2n=32.

20.4 Morphology

***20.4.1 Habit*:** A graceful, erect and unbranched palm; height depends on cultivars and environmental conditions.

***20.4.2 Root*:** Adventitious root system, typical of monocots produced from base bole. Primary root branches to give secondary and tertiary roots. Maximum concentration of roots is within a radius of 1 m from base and usually in the top 60 cm of soil. Root hairs are absent.

***20.4.3 Stem*:** Stem becomes visible when palm is 3 years old. Girth of stem depends upon genetic variation and soil conditions. Mature stem is single, cylindrical, 30 m tall and 25-40 cm in diameter. It is green when young becoming greyish brown with age and ringed with leaf scars. Internodal distance reduces with age and stem becomes thinner under unfavourable conditions.

***20.4.4 Leaf*:** Leaves are borne in terminal crown about 2.5 m in diameter with phyllotaxy 2/5. Adult leaves are paripinnate with long smooth sheathing base completely encircling stem. 1 year old seedling has 4-5 leaves, which increases to 8-12 in adult palm. Life of an unfurled leaf is 2 years. Six new leaves are produced per year.

***20.4.5 Inflorescence*:** Flowering begins when 4-6 years old. First inflorescence appears at 10th node at a height of 1.5-2 m from ground, produced in axil of every leaf, leaf sheath closely covering it until opening, usually 3-4 inflorescences per year. Inflorescence is a spadix, male and female flowers on the same spadix (monoecious), boat shaped spathe enclosing the spadix, spadix branched to secondary (12-16) and tertiary rachis bearing female flowers (up to 600) on thickened bases and male flowers (15,000- 50,000) arranged in pairs of two rows along upper part of thin branches.

20.4.6 Flower:

20.4.6.1 Male flowers: Sessile, creamy white, triangular.

Perianth: 6 tepals arranged in two whorls, imbricate aestivation.

Androecium: 6 stamens.

20.4.6.2 Female flowers: Sessile, cream coloured turning to green.

Perianth: 6 tepals arranged in two whorls.

Gynoecium: Bicarpellary, syncarpous, unilocular, single ovule in basal placentation, dome shaped trifid stigma formed by 3 stiff stylar projections.

Fruit: Ovoid drupe with fibrous pericarp, orange red to scarlet in colour when ripe, size and shape variable, 100-125 fruits/spadix.

Seed: Ruminate endosperm

20.5 Floral Biology

Male flowers open on same day or a few days after spathe bursts open exposing spadix. Individual flowers start opening from tip downwards. Anther dehisces simultaneously with opening of flowers. Male flowers drop off either on the same day or the following morning. After male flowers are shed, clear nectar oozes out from point of attachment of male flowers. Male phase in arecanut lasts for 25-46 days.

Female flower buds at the time of opening of spathe are cream coloured which turn to green. They start opening after all male flowers are shed. Anthesis is from 2 AM to 10 AM. Female phase extends for 3-10 days. Maximum stigmatic receptivity is between 2nd and 4th day of opening. Middle aged palms have higher stigmatic receptivity than young and old palms.

20.6 Pollination

General mode of pollination is cross-pollination. Overlapping of male and female phases leads to self-pollination. Wind is the main agent of pollination and pollen is carried up to 1-2 km.

20.7 Emasculation and Crossing Techniques

Remove portion of rachillae bearing male flowers soon after the inflorescence frees itself from spathe. Cover the inflorescence with female flowers with a cloth bag. Anther from desired male parent is rubbed against receptive stigma or pollen is dusted on the stigmatic surface. The bag is replaced immediately after pollination and the process is repeated till all the female flowers in the inflorescence open.

20.8 Breeding Objectives

- To develop cultivars with high yield.
- To breed cultivars with regular bearing.
- To develop cultivars with large sized fruits, more number of nuts/bunch.
- To develop cultivars with good quality.
- To develop cultivars with more number of female flowers and high percentage of seed set.

- To breed dwarf cultivars.
- To develop cultivars with tolerance to yellow leaf disease.

20.9 Breeding Methods

20.9.1 Introduction

In arecanut, improved cultivars have been evolved through introduction of indigenous and exotic types and refinement of selection procedure for mother palm, seed nut and seedling. This has resulted in the development of many improved cultivars, e.g., Mangala, Sumangala, Sreemangala, Mohitnagar, SAS-1 etc.

20.9.2 Hybridization

Hybridization programme in arecanut was started in early 1970s at the CPCRI Regional Station, Vittal. Inter-varietal hybridization carried out among Mangala, Sumangala, Sreemangala, Mohitnagar, Thirthahalli, Sreevardhan and Hirehalli Dwarf and evaluation of seedlings with respect to their performance led to the identification of two hybrids, namely, VTLAH-1 and VTLAH-2.

In interspecific hybridization between *Areca catechu* and *Areca triandra,* the hybrid expressed hybrid vigour for number of female flowers, length of spadix and girth of stem but showed high sterility.

20.10 Cultivars

Mangala (VTL 3): An introduction from South China; semi-tall, early bearing, more female flowers/inflorescence, high nut set, average yield 8.25 kg ripe nuts/palm/year.

Sumangala (VTL 11): An introduction from Indonesia; tall, nuts deep yellow to orange, average yield 12.93 kg ripe nuts/palm/year at the age of 10 years.

Sreemangala (VTL 17): An introduction from Singapore; tall, ripe nuts deep yellow colour, average yield 12.82 kg ripe nuts/palm/year.

Mohitnagar: An indigenous cultivar from West Bengal; known for uniformity, wide adaptability, high yield of 15 kg ripe nuts/palm/year.

SAS-1: A selection from Sirsi (Karnataka) area; tall, regular bearing, branch compact, average yield 18 kg ripe nuts/palm/years; suitable for tender as well as ripe nut processing.

VTLAH-1 **(Vittal Arecanut Hybrid-1):** Hybrid between Hirehalli Dwarf × Sumangala; dwarf, sturdy stem, super imposed nodes, reduced canopy size, well spread leaves, partial drooping crown, medium size, oval to round and yellow orange-coloured nuts, high recovering of chali (26-45%), average chali yield 2.55 kg/palm/year.

VTLAH-2 (Vittal Arecanut Hybrid-2): Hybrid between Hirehalli Dwarf × Mohitnagar; dwarf, reduced canopy, early yield stabilization, oval, medium, orange nut, average yield (chali/palm/year) 2.64 kg.

Review Questions

Part A

Answer the following questions true or false.

1. The chromosome number of arecanut is 2n=32. True/False
2. General mode of pollination in arecanut is self-pollination. True/False
3. The perianth of arecanut contains 3 tepals. True/False
4. SAS-1 is a hybrid. True/False
5. Wind is the main agent of pollination in arecanut. True/False

Part B

Write a brief note on each of the following.

1. Breeding methods in arecaanut.
2. Breeding objectives of arecanut.
3. Floral biology of arecanut.
4. Cultivars of arecanut.
5. Emasculation and crossing techniques in arecanut.

21

Breeding of Cashewnut

21.1 Introduction

Cashewnut or cashew is an export-oriented crop grown for its nuts, a true drupe. It is often referred to as 'wonder nut'. It is highly delicious and nutritive. Its nutritional composition compares well with almond, hazelnut and walnut. The economic part is kernel which is a unique combination of fats (47%), protein (21%), carbohydrates, minerals and vitamins. Nutritionally it is equivalent to milk, egg and meat, it is good appetizer, an excellent nerve tonic, stimulant, body builder. India is the largest producer, processor and exporter; and second largest consumer in the world. Cashew is grown in Kerala, Karnataka, Goa and Maharashtra along the west coast and in Tamil Nadu, Andhra Pradesh, Odisha and West Bengal along the east coast.

21.2 Origin and Distribution

Cashewnut is believed to be a native of Brazil from where it was distributed to many tropical areas.

21.3 Taxonomy and Classification

The taxonomic position of cashewnut as per Hutchinson is:

Phylum	:	Angiospermae
Sub-phylum	:	Dicotyledones
Division	:	Lignosae
Order	:	Sapindales
Family	:	Anacardiaceae
Genus	:	*Anacardium*
Species	:	*occidentale*
Binomial name	:	*Anacardium occidentale* L.

Twenty species of *Anacardium* are known but only *A. occidentale* is commercially cultivated in all the countries for the edible kernels.

The chromosome number of cashewnut (*A. occidentale*) is 2n=2x=42.

21.4 Morphology

***21.4.1 Habit*:** Spreading, evergreen tree, height up to 12 m.

***21.4.2 Root*:** Extensive lateral root system and a deep taproot.

***21.4.3 Stem*:** Erect, canopy conical or umbrella shaped (when grown under favourable conditions and unharmed by pests). Under less favourable conditions tree is much smaller, stem tortuous and branches rest on ground creeping over soil and rooting takes place where they touch soil.

***21.4.4 Leaf*:** Petiole swollen at base and flattened on upper surface, alternate phyllotaxy, simple, glabrous, leathery, oblong to obovate, rounded apex; veins prominent, 10-20 pairs, young leaves reddish brown later turning dark green after 20 days.

***21.4.5 Inflorescence*:** Terminal panicle, conical or pyramidal in shape, containing 120-1100 flowers.

***21.4.6 Flower*:** Bracteate, scented, white in colour at first soon turning to pink after a few days, andromonoecious with male and bisexual flowers in the same inflorescence.

21.4.7 **Calyx:** 5 sepals.

***21.4.8 Corolla*:** 5 petals.

***21.4.9 Male flower*:**

Androecium: 10 stamens, 9 short and one long with red anther projecting above corolla.

21.4.10 Bisexual flower

Androecium: 10 stamens, 9 short and one long projecting just above corolla but below stigma.

Gynoecium: Ovary unilocular with single ovule, style simple and exerted.

Fruit: Kidney shaped fruit called drupe, greyish-brown in colour with a hard pericarp consisting of a leathery exocarp, spongy mesocarp containing cashew nut shell liquid and a hard and brittle endocarp. Cashew apple, a false fruit, is the swollen pedicel.

Seed: Seed (kernel) kidney shaped with reddish brown testa, 2 large, white cotyledons and a small embryo.

21.5 Floral Biology and Pollination

Peak anthesis is between 9 AM and 11 AM. Stigma is receptive as soon as flower opens and remains receptive for 48 hours from anthesis. Anther dehiscence takes place 1-5 hours after anthesis. Pollination takes place through bees which transfer sticky pollen to stigma. In bisexual flower the stamen is shorter than stigma leading to cross-pollination.

21.6 Emasculation and Crossing Techniques

- Panicles having flower buds that will open the next day are selected on both female and male parental trees. Remove the opened flowers.
- Everyday morning (8:30-9:30AM) remove all the opened female flowers on the selected panicle on female parental trees. The anthers are removed using an ordinary pin before dehiscence.
- Freshly opened male flowers with indehiscent anthers are collected in a petri-dish from a selected male parent only and anthers are allowed to dehisce under partial shade.
- Butter paper roll from emasculated flower is removed. Pollen of a male parent is collected in the petridish.
- The pollinated stigma is re-enclosed in butter paper roll along with style.
- Each panicle is labelled indicating names of female and male parents of the cross and also panicle number.
- Each panicle is used for only one cross combination. The same procedure is continued for 8-10 bisexual flowers.
- All the opened bisexual flowers that are not used in pollination are removed daily.
- All the remaining flower buds are removed from the panicle.
- Each panicle with a developing hybrid nut is enclosed in a cloth bag in order to collect panicle wise. The details of the cross should also be written on the cloth bag.
- The hybrid seedlings are raised in polybags.

21.7 Breeding Objectives

- To evolve cultivars having high yield potential with good quality nuts.
- To develop cultivars having dwarf and compact canopy with intensive branching habit.
- To evolve cultivars with bold kernel to fetch premium prices in international market.
- To breed cultivars tolerant/resistant to biotic stresses (tea mosquito, stem and root borer, thrips, anthracnose, etc.) and abiotic stress (drought).
- To develop suitable rootstock having dwarfness, stress resistance and easily asexually propagated.

21.8 Breeding Methods

21.8.1 Introduction

Cashew was introduced into India in Goa region and Malabar Coast during 16th century. However, due to small quantity seed/tree, there was a limited genetic base from which all the present-day cultivars in the country have been developed. This may be the reason for low variability in the cultivated cashew in India.

21.8.2 Selection

Survey was undertaken by various centres in India to identify and select elite cashew plant. Cultivars developed through selection are:

Vengurla-1, BLA-139-1, K-19-1, K-10-2, K-22-1, K-10-1, K-16-1, T-1, Murude ST-94, Seed Farm Selection-4, BPP-241 and Goa-1.

21.8.3 Hybridization

It is one of the potential tools to evolve new cultivar. Cashewnut is highly heterozygous plant due to cross-pollination. Examples of hybrids are:

V-3, V-4, V-5, V-6, BPP-1, BPP-2, BPP-8, Priyanka.

21.8.4 Mutation Breeding

No improved cultivar has so far been developed through mutation breeding. Work initiated at the Regional Fruit Research Station, Vengurla showed that the irradiation of cashewnut bud sticks at 3 kr gamma rays dose led to mortality of all the bud sticks.

21.9 Cultivars

Cashew cultivars recommended for different states are given below:

Andhra Pradesh: BPP-4, BPP-6, BPP-8, VRI-2, Madakathara-1.

Andaman and Nicobar Islands: Ullal-1, VRI-2, V-1, V-4.

Chhattisgarh: V-4, Madakathara-1.

Goa: V-1, V-4, V-6, VRI-2, Goa-1.

Karnataka: NRCC Sel-1, NRCC Sel-2, Ullal-1, Ullal-2, VRI-1, VRI-2, V-1, V-4.

Kerala: Madakathara-1, Madakathara-2, K-22-1, Dhana.

Maharashtra: V-1, V-4, V-6, VRI-2.

Manipur: VRI-2, Madakathara-1.

Odisha: Bhubaneswar-1, VRI-2, BPP-2.

Tamil Nadu: VRI-1, VRI-2, VRI-3.

West Bengal: Jhargram-1, Madakathara-1.

Review Questions

Part A

Answer the following questions true or false.

1. Cashewnut is believed to be a native of Mexico. True/False
2. Peak anthesis of cashewnut is between 8 AM and 9 AM. True/False
3. Cashew was introduced into India during 16th century. True/False
4. The chromosome number of *A. occidentale* is 2n=52. True/False
5. Anther dehiscence takes place 1-5 hours after anthesis in cashewnut. True/False

Part B

Write a brief note on each of the following.

1. Emasculation and crossing techniques in cashewnut.
2. Breeding objectives of cashewnut.
3. Morphology of cashewnut.

22

Breeding of Tea

22.1 Introduction

Tea is the most important non-alcoholic beverage gaining further popularity as an important "health drink". It is consumed as a morning drink by 2/3rd of world population. Tea was initially used as a medicine and subsequently as beverage and now has proven to be a potential raw material for pharmaceutical industry. Tea is mainly consumed in the form of 'fermented tea' or 'black tea', 'non-fermented tea' or 'green tea' and 'semi-fermented tea' or 'oolong tea'. Green leaves are also used as vegetables such as 'leppet tea' in Burma and 'meing tea' in Thailand. Tea leaves have more than 700 chemical constituents among which flavonoids, amino acids, vitamins (C, E, K), caffeine and polysaccharides are important to human health. The stimulating effect of tea is due to caffeine (1.25-4.5%). Per cup value of caffeine in tea is three and half times less than that of coffee and is less harmful than coffee. Importantly, the vitamin C content in leaves is comparable to that of lemon.

22.2 Origin and Distribution

The centre of origin of tea is South East Asia. The cultivation of tea in India dates back to the introduction of tea seeds and plants of the China type (*Camellia sinensis*) from China in the beginning of the 19th century when tea gardens were started in Assam, the Himalayas and the Nilgiris. The Assam type (*C. assamica*), which was superior to the China type, paved the way for the commercial exploitation of tea. A third type, Cambod (*C. assamica* ssp. *lasiocalyx*) was introduced in India in the beginning of the 20th century from South East Asia. These species growing side by side crossed freely among themselves making early tea population highly heterogeneous.

22.3 Taxonomy and Classification

The taxonomic position of tea as per Hutchinson is:

Phylum : Angiospermae

Sub-phylum : Dicotyledones

Division : Lignosae

Order : Theales

Family : Theaceae (syn. Camelliaceae)

Genus : *Camellia*

Species : *sinensis*

Binomial name : *Camellia sinensis* (L.) Kuntze

The genus *Camellia* comprises 82 species (Sealy, 1958) and accounts for more than 325 species in 2000 (Mondal, 2002) indicating genetic instability and high out-breeding nature of the genus. Tea breeds well with wild relatives and suspected involvement of species in tea gene pool was investigated. Two particularly interesting taxa are *C. irrawadiensis* and *C. taliensis* whose morphological distributions overlap with that of tea. It is postulated that some desirable traits such as anthocyanin pigmentation or special quality character of Darjeeling tea might have been introduced from wild species. Other species of *Camellia* suspected to have contributed to tea gene pool by hybridization include *C. flava, C. petelotii* and possibly *C. lutescens.* Present day commercial tea populations are thus polymorphic in origin resulting in wide genetic diversity. As botanical classification of species of *Camellia* is a subject of intense speculation, tea is generally referred as *Camellia sinensis,* irrespective of species-specific differences. Botanists distinguished three distinct tea-producing taxa, which were referred as 'jats'. They are given below in brief:

i) **China type** (*Camellia sinensis*): Shrub, 1-3 m height, erect branches; leaves thick, small, deep green and erect; 2 morphological forms: macrophylla (broad and long leaves) and parviflora (small and narrow leaves), both forms with elliptic leaf base, obtuse apex and markedly serrate margin; resistant to cold and adverse conditions; low yielding; flowers borne singly.

ii) **Assam type** (*Camellia assamica*): Small tree,10-15 m height; leaves drooping, large, glossy, acuminate apex, distinct marginal veins, leaf blade broadly elliptic, indistinctly denticulate margin; 2 types: Assam type (light green leaves, higher yield, better quality tea) and Manipuri type (dark green leaves, drought resistant, poor yield, poor quality tea); flowers in clusters of 2-9; adapted to tropical conditions.

iii) **Cambod type** (*Camellia assamica* ssp *lasiocalyx*): Conical appearance, 6-10 m height; leaves semi-erect, vary in size between China and Assam types, elliptic, marginal veins not prominent.

The chromosome number of *Camellia* is 2n=2x=30.

22.4 Morphology (*C. sinensis*)

***22.4.1 Habit*:** Under natural conditions, small tree about 9 m height; under cultivation, pruned to 0.5-1m.

***22.4.2 Root*:** Strong taproot and lateral roots giving rise to surface mat of feeding roots confined to 60 cm surface layers, starch stored in roots.

***22.4.3 Stem*:** Two different types of shoots: aperiodic shoot and periodic shoot; aperiodic shoot arises from pruned bush frame and growth is continuous; periodic shoot develops from axillary bud of aperiodic shoots and shows rhythmic growth. Term 'flush' describes growth by apical bud between two successive phases of dormancy. In periodic shoots, the bud passes through alternate phase of growth and dormancy. Dormancy phase is called "banji" phase.

***22.4.4 Leaf*:** Evergreen, obovate to lanceolate, simple, alternate, glossy on upper surface and sparsely hairy on lower surface; margin serrate.

***22.4.5 Flower*:** Solitary or in clusters of 2-4 with short peduncles in axils of scale leaves on current season's growth. Sepals and petals 5-7; numerous stamens; ovary superior, 2-4 loculed.

***22.4.6 Fruit*:** Thick-walled capsule, taking 9-12 months to mature.

22.5 Pollination

Tea is entirely cross-pollinated. Pollination is carried out by insects. Tea is said to be 'virtually sterile' as selfing gives a much lower percentage of viable seeds. Occurrence of self-incompatibility in certain Assam jats which is lacking in Chinese or Chinese-Assam hybrids was reported.

22.6 Breeding Objectives

- To develop cultivars superior in yield and quality.
- To develop cultivars with resistance to drought and major pests and diseases.

22.7 Breeding Methods

22.7.1 Clonal Selection

Tea population is highly heterogeneous and it can be propagated vegetatively. Hence, clonal selection programme was initiated at Tocklai Experimental Station, Jorhat, Assam and UPASI Tea Research Institute, Valparai, Coimbatore, Tamil Nadu. This has resulted in the identification of certain improved clones. Clonal selection is based on bush morphology, yield, crop quality, tolerance to biotic and abiotic stresses, rooting behaviour, etc. A total of 197 clones have been released so far and some of the important clones are:

UPASI-2 (Jayaram), UPASI-3 (Sundaram), UPASI-6 (Brooklands), UPASI-8 (Golconda), UPASI-9, UPASI-10 (Pandian), UPASI-14 (Singara), UPASI-17, TRI 2024 (Sri Lanka), TRI 2025 (Sri Lanka).

22.7.2 Hybridization

Natural Hybrids: Natural hybrids between Assam and China types were earlier in common use. They were designated either 'China hybrids' or 'Assam hybrids' depending upon their morphological closeness to parents. They had their own limitations for both yield and quality, though in some cases both the traits were of high orders.

Polyclonal Seed Stock: A group of selected clones are planted in isolated polycross nursery called 'polyclonal seed bari' and allowed to freely cross among themselves. Such seeds are called 'polycross seeds' or 'polyclonal seed stock'. For the first time in India, a seed cultivar named 'Stock 203' (Gauri Shankar) was released in 1954. However, this approach was discontinued due to uncertain pollination pattern and field performance of such polyclonal stocks.

Biclonal Seed Cultivar: Biclonal seed cultivar, i.e., hybridization between a pair of desirable clones is resorted to produce seed cultivars. Biclonal hybrid progenies show a fair amount of hybrid vigour and morphological uniformity. The problem of this breeding technique is the long period (about 23-27 years) to develop a seed cultivar. However, in the modified method it takes 10-12 years only. Some of the important biclonal stocks released are TS-397, TS-449 and TS-450.

Interspecific Hybridization: Interspecific hybridization programme at Tocklai was initiated to study the species relationship and produce heterotic hybrids. Tea was crossed with *C. irrawadiensis*. Its hybrids with high vigour were studied. A few selected hybrid clones were backcrossed with high quality

Assam clones. On evaluation of backcrossed hybrid progenies, a high yielding, high quality clone 'TV-24' was selected and released.

Interspecific hybridization between *C. sinensis* and *C. japonica* was successful. The hybrid is a triploid (2n=45). The growing habit, leaf and flower characteristics are intermediate between the two parents. The tea cultivar used in this cross was a tetraploid (2n=60) and *C. japonica* was a diploid (2n=30).

22.7.3 Polyploidy Breeding

Open-pollinated progenies often produce natural diploid, triploids, tetraploids and aneuploids. But tetraploids are not of very high quality. Therefore, attempts were made more towards production of high-quality triploids by controlled hybridization between natural tetraploids and high-quality diploids. The triploids thus produced were superior to normal diploids in their rooting ability, leaf size and dry weight. So, breeding may have to be concentrated mostly on production of vigorous triploids (2n=45) provided the quality aspects do not suffer.

Out of 56 triploid progenies evaluated for their rooting behaviour, yield, crop quality and tolerance to pests and diseases at Tocklai, five were identified as promising out of which one was released as clone TV-29 in 1990. UPASI-3 (Sundaram) and UPASI-20, selected from natural variations in South India, were found to be triploid and were released much earlier than TV-29.

22.7.4 Mutation Breeding

No natural mutants have been recorded nor has any cultivar been bred through this technique. Tea tolerates 2-6 kr of gamma radiation. Mutants produced by irradiation of cuttings and seeds had reduced vigour, stunted growth and less number of branches and foliage. The use of irradiated pollens also caused fruit drop.

Review Questions

Part A

Answer the following questions true or false.

1. The centre of origin of tea is South West Asia. True/False
2. Tea population is highly heterogeneous. True/False
3. Interspecific hybridization between *C. sinensis* and *C. japonica* was not successful. True/False

4. Tea is entirely cross-pollinated. True/False
5. No natural mutants have been recorded. True/False

Part B

Write a brief note on each of the following.

1. Polyploidy breeding in tea.
2. Hybridization in tea.
3. Clonal selection in tea.
4. Taxonomy and classification of tea.
5. Origin and distribution of tea.

23

Breeding of Coffee

23.1 Introduction

Coffee is the second important popular beverage for its aroma and mild stimulant action. Its dried beans are roasted, ground and brewed to make a stimulating and refreshing beverage. Use of coffee has evolved from original chewing of leaves and beans of the plant to relieve pain, hunger and fatigue to the present more sophisticated uses, like espresso and decaffeinated coffees. Its use was first discovered in Arabia around mid-15th century. Nearly 75% of world's coffee is produced from *Coffea arabica,* 24% from *C. canephora* and 1% from *C. liberica.* Arabica coffee is known for its aroma and low caffeine content (1-1.5%), Robusta coffee characterized by high caffeine content (2-2.5%) is preferred for manufacture of instant coffee. Liberica coffee with a bitter taste is used as filler with other coffees. Habitual consumption of coffee is associated with substantially lower risk of type-II diabetes. Caffeine and another component theophylline are strong stimulants of pancreatic cells. In Ethiopia, dried coffee berries are used as masticatory since ancient times. Ground roasted coffee is also mixed with fat and eaten. Coffee pulp and parchment are used as manures and mulches. In India, they are occasionally fed to cattle. Caffelite, a type of plastic, can be made from coffee beans.

23.2 Origin and Distribution

Majority of *Coffea* species are native to Africa. *Coffea arabica* is native to Ethiopia. *Coffea canephora* is a native of Central Africa (Congo and Zaire). Coffee was introduced into India in 1600 AD by a Muslim pilgrim, Baba Budan.

23.3 Taxonomy and Classification

The taxonomic position of coffee as per Hutchinson is:

Phylum	:	Angiospermae
Sub-phylum	:	Dicotyledones
Division	:	Lignosae
Order	:	Rubiales
Family	:	Rubiaceae
Genus	:	*Coffea*
Species	:	*arabica, canephora, liberica*
Binomial name	:	*Coffea arabica*
		Coffea canephora
		Coffea liberica

There are about 70 species of *Coffea.* They are grouped into 3 sections: Eucoffea, Mascarocoffea and Paracoffea (more correctly as Coffea). Argocoffea is excluded from Coffea since seeds do not resemble those of Coffea.

Section Coffea includes most of useful species of the genus. Based on factors such as height of tree, leaf thickness, fruit colour and geographical distribution, it is further sub-divided into 5 sub-sections, namely, Erythrocoffea, Nanocoffea, Pachycoffea, Melanocoffea and Mozambicoffea. *Coffea arabica* and *Coffea canephora* (Erythrocoffea) and *Coffea liberica* (Pachycoffea) are few of the species that are cultivated commercially in India. *C. arabica*, *C. canephora* and *C. liberica* are commonly known as arabica coffee, robusta coffee and tree coffee respectively.

Majority of the species are diploid (2n=2x=22) but *C. arabica* is a tetraploid (2n=4x=44).

23.4 Morphology

***23.4.1 Habit*:** Evergreen, shrub or small tree, up to 5 m height when pruned.

***23.4.2 Root*:** Short, stout taproot rarely extending beyond 45 cm; 4-8 axial roots originating as laterals from taproot going down vertically to 2-3 m or more; many lateral roots, 1-2 m long in a horizontal plane, below them lower laterals ramify evenly and more deeply in the soil.

***23.4.3 Stem*:** Dimorphic branching due to differential development of two buds, which occur one above other in each leaf axil of central orthotropic stem; under the dominance of apical bud, only upper bud in leaf axil grows to produce plagiotropic lateral or primary branch arising on opposite side of each node in succession from base upward around stem; when main stem is topped or damaged, lower bud grows to produce an orthotropic shoot called sucker or water shoot.

***23.4.4 Leaf*:** Petiolate, stipulate, opposite, decussate in suckers but in plagiotropic branches, successive nodes with leaves lie in one plane, dark green when mature, young leaves bronze-tipped in Arabica, elliptical, apex acuminate, margin sometimes undulate, domatia or small cavities on lower surface at insertion of lateral veins giving slight protuberance on upper surface.

***23.4.5 Inflorescence*:** Condensed cyme arising in leaf axils of mature wood, on short peduncles and subtended by bracts; 4-5 inflorescences of 1-4 flowers each in Arabica and 5-6 flowers each in Robusta per axil.

***23.4.6 Flower*:** Fragrant, white, pentamerous.

***23.4.7 Calyx*:** 5 sepals, small, cup-shaped, rudimentary, gamosepalous, valvate aestivation.

***23.4.8 Corolla*:** 5 petals, tubular, gamopetalous, imbricate aestivation.

***23.4.9 Androecium*:** 5 stamens, epipetalous inserted on corolla tube between lobes.

***23.4.10 Gynoecium*:** Ovary inferior and bilocular with one ovule in each locule, stigma bifid.

***23.4.11 Fruit*:** Drupe, containing 2 seeds normally, oval, elliptic, green when immature and on ripening yellow and then crimson; exocarp (outer skin) smooth and tough, mesocarp (pulp) soft, yellowish; endocarp (parchment) greyish green, fibrous made up of stone cell or scleroides surrounding seeds.

***23.4.12 Seed*:** Ellipsoidal, pressed together by flattened surface which is deeply grooved and outer-surface is convex; green endosperm; small embryo near the base, testa (silver skin) thin, silvery.

23.5 Pollination

Pollination takes place 6 hours after anthesis under bright light and warm windy conditions. Wind, gravity and bees are agents of pollination. In *C. arabica,* flowers have short corolla tube, long style and exerted stamens which promote self-pollination. Arabica coffee is self-pollinated with different degrees of natural out-crossing. *C. canephora* is cross-pollinated due to self-incompatibility.

23.6 Breeding Objectives

- To develop high yielding lines having moderate to high rust resistance, bold bean size and with good cup quality.
- To develop lines with drought tolerance, low caffeine content.

23.7 Breeding Methods

Central Coffee Research Institute (CCRI), Balehanur, Karnataka is the only organization in India which takes up crop improvement work. As a result, a number of superior selections in Arabica and Robusta coffee have been released.

23.7.1 Introduction

Many breeding lines and hybrids were introduced to CCRI, Balehanur. Hibrido-de-Timor (HDT), a natural hybrid of robusta and arabica origin, spotted in Timor Island was introduced into India in 1961. San Raman, a dwarf mutant of arabica, spotted in Costa Rica was introduced into India to develop dwarf cultivars.

23.7.2 Mass Selection

Mass selection is useful in the initial selection of superior genotypes in cross-pollinated robusta coffee (*C. canephora*). Robusta selections S.270 and S.274 were derived from the eight superior mother plants selected from indigenous collections.

23.7.3 Clonal Selection

Among the descendant families of S.270 and S.274, seventeen mother plants were identified to be superior as they yielded twice the mean of the family. They were multiplied clonally. The clones derived from these mother plants were better than their seedling counterparts. These clones are known as Balehanur Robustas.

23.7.4 Hybridization

Many natural hybrids between arabica and robusta coffee (e.g., HDT) have been identified. Besides, hybridization between cultivars and back-crossing has been taken up extensively to improve yield, disease resistance and bean quality. Examples: S.795, Selection 7, Selection 9, Selection 10.

23.7.5 Interspecific Hybridization

Selection 16 is an interspecific hybrid of Robusta × Arabica origin evolved at CCRI. Its performance has not been consistent in many of the estates.

Selection 11 is an interspecific hybrid of *C. liberica* × *C. eugenioides* origin with tetraploid status and characters resembling arabica and high resistance to rust. In order to improve the bean size, it was crossed with different selections and eight crosses were established. Two of such crosses, namely, S 4595 and 4600 are showing improved bean size and cup quality, while maintaining high rust resistance.

S.274 robusta is a high yielder with bold bean character and *congensis* is a species showing compact bush size, bean quality better than robusta but is a poor yielder. By crossing these two, the compact bush size with bold bean character, satisfactory yield and cup quality has been achieved. This interspecific hybrid is known as Congensis × Robusta (C×R).

23.7.6 Mutation Breeding

Only certain natural mutants have been spotted out. No cultivar has been released through mutation breeding.

23.8 Cultivars

Arabica Cultivars: Selection 1 (S.288), Selection 3 (S.795), Selection 5, Selection 6, Selection 7, Selection 8, Selection 9, Selection 10 (Caturra crosses), Selection 11, Cauvery, Chandragiri coffee.

Robusta Cultivars: Sel-1R (S-274), Sel-2R (S-270), Sel-3R.

Review Questions

Part A

Answer the following questions true or false.

1. Majority of *Coffea* species are native to Africa. True/False
2. Coffee was introduced into India in 1900 AD. True/False
3. *C. canephora* is cross-pollinated. True/False
4. Nearly 40% of world's coffee is produced from *Coffea arabica.* True/False
5. Selection 11 is an interspecific hybrid. True/False

Part B

Write a brief note on each of the following.

1. Breeding methods in coffee.
2. Morphology of stem and leaf of coffee.
3. Cultivars of coffee.

24

Breeding of Oil Palm

24.1 Introduction

In India, the commercial cultivation of oil palm started during 1970. Today about 16 countries of tropics cultivate oil palm. The palm oil has been safe and nutritious source of edible oil for human consumption for 1000 years. Palm oil and its liquid fraction palmolein is consumed worldwide as cooking oil. Malaysia, Indonesia, Nigeria and Columbia started to grow oil palm in a large scale during 1980. The commercial producing countries are Liberia, Peru, Angola, Senegal, Ghana, etc.

24.2 Origin and Distribution

Oil palm is native to West Africa. It has been first introduced to Malaysia during 18th century. During 1960s it has been introduced to the Philippines.

24.3 Taxonomy and Classification

The taxonomic position of oil palm as per Hutchinson is:

Phylum	:	Angiospermae
Sub-phylum	:	Monocotyledones
Division	:	Corolliferae
Order	:	Arecales (syn. Palmales)
Family	:	Arecaceae (syn. Palmae)
Genus	:	*Elaeis*
Species	:	*guineensis*
Binomial name	:	*Elaeis guineensis* Jacq.

Table 24.1 Features differentiating fruit types of oil palm

Sl No.	Characters/ Composition	Dura	Tenera	Pisifera
1.	Mesocarp proportion in fruit (%)	35-50	60-96	98
2.	Shell thickness (%)	2-8	0.5-4	-
3.	Oil percentage	15	36	25
4.	Average proportion of shell in fruit (%)	30	10	-
5.	Average proportion of kernel in fruit (%)	16	16	10

The commercial oil palm is hybrid between Dura × Pisifera and they are called Tenera type. Hence, most of the cultivars have been developed within a population of dura and pisifera. The germplasm of dura and pisifera are the parental population used in the hybrid production programme. The other common breeding population are Deli, Avros, Yangambi, La me, Binga, Ekona, Calabar, Deli × Avros, Deli × Yangambi.

Oil palm is a diploid with 2n=2x=32.

24.4 Morphology

24.4.1 Habit: Tall, unbranched, 30 m height with a crown of 4-5 opened palmate leaves.

24.4.2 Root: Root system consisting of primary root which are either orthogravitropic or diagravitropic.

24.4.3 Stem: Solitary columnar stem, short internodes.

24.4.4 Leaf: Plant produces two leaves/month, a crown of 4-5 opened palmate leaves, irregular set of leaflets on the leaf.

24.4.5 Inflorescence and Flowers: Compound spike or spadix; stout peduncle of 30-45 cm long; central rachis with spirally arranged spikelets of 100-200 in number in different palms; two spathes enclose the inflorescence.

Male inflorescence finger-like, spineless cylindrical spikes, 10-20 cm long each with 700-1200 closely packed small male flowers sunk in tissues of rachis. Each flower has a triangular bract, perianth of 6 minute segments, tubular androecium with 6 anthers and rudimentary gynoecium with 3 projections corresponding to trilobed stigma.

Female inflorescence has thick and fleshy spikelets in axils of spiny bract. Flowers are arranged spirally in shallow cavities in rachis with 12-30 flowers on central spikelets and 12 or less in lower and upper spikelets. Each female flower is usually with a small male flower on either side, which normally aborts. Female flower has 2 bracteoles, 6 sepaloid tepals in two whorls in perianth, rudimentary androecium of 6-10 short projections, tricarpellary ovary of which one carpel develops and sessile trilobed stigma.

***24.4.6 Fruit*:** Sessile drupe, pericarp — exocarp (skin), mesocarp (pulp), endocarp (seed), surrounding kernel — which is testa/skin having solid endosperm, embryo.

24.5 Floral Biology and Pollination

Flowering begins early when young palm is well-established. An inflorescence primordium is produced in axil of each leaf at the time of leaf initiation. Inflorescence reaches central spear stage in two years and a further 9-10 months are required for flowering and anthesis. Each flower primordium is a potential producer of male and female organs but one or other usually remains rudimentary to produce either a male or female inflorescence. Hermaphrodite flowers and inflorescence are occasionally produced. Length of male and female period varies widely but period of 4-5 months may predominate, in which 8-10 inflorescences of one sex are produced. In a plantation, some palms must produce male inflorescence and shed pollens at same time as others have female inflorescences with receptive flowers so that pollination may take place. Palm is thus monoecious with a tendency towards dioecious habit. The period between sex differentiation and anthesis is about 2 years. Oil palm is a cross-pollinated crop. Wind is the sole agent of pollination.

24.6 Breeding Objectives

- High yield of fresh fruit bunches, fresh fruit bunch oil and kernel.
- High quality oil with high level of unsaturation, high iodine value, high vitamin -E and carotenoids.
- Desirable tree structure with reduced trunk growth and long bunch stalks for easy harvest.
- Breeding for compact palm, to allow high planting density of 180 palms/ha.

24.7 Breeding Methods

24.7.1 Selection

Studies so far have indicated that nursery selection is unlikely to result in substantial genetic improvement.

24.7.2 Hybridization

In India, hybridization work on oil palm was initiated at the Research Centre at Palode, Kerala in 1975 using Dura parents of Malaysian origin with four pollen samples imported from Nigeria. Eleven such Dura × Pisifera combinations were planted in 1976 at Palode. Crosses 65 D × 30.103 P, and 120 D × 30.103 P performed much better than others. These crosses have shown potential of giving 45 tonnes of oil/hectare/year. The two hybrids were recommended for release as Palode I and Palode II at XIII AIRCP on Palm at Jorhat in February 1998.

Review Questions

Part A

Answer the following questions true or false.

1. In India, the commercial cultivation of oil palm started during 1970. True/False
2. Oil palm is native to East Africa. True/False
3. Oil palm is a diploid with 2n=2x=32. True/False
4. Oil palm is monoecious with a tendency towards dioecious habit. True/False
5. In India, hybridization work on oil palm was initiated in 1970. True/False

Part B

Write a brief note on each of the following.

1. Floral biology and pollination in oil palm.
2. Breeding objectives of oil palm.
3. Hybridization in oil palm in India.

Bibliography

Abraham Z 2017. *Fruit Breeding*. Agri Horti Press, New Delhi.

Bhandari MM 1979. *Practicals in Plant Breeding* (A manual-cum-practical record), 2nd ed. Oxford & IBH Publishing Co, New Delhi, pp 1-98.

Chaudhari SM and Desai UT 1993. Effects of plant growth regulators on flower sex in pomegranate (*Punica granatum* L.). *Indian J. Agr. Sci.*, 63:34–35.

Dinesh MR 2015. *Fruit Breeding*. New India Publishing Agency, New Delhi.

Dinesh MR, Ravishankar KV and Sangma, D 2016. Mango breeding in India - Past and future. *J. Hort. Sci.*, 11 (1): 1-12.

Dutta AC 1974. *Botany for Degree Students*, 4th ed, Reprint 1976. Oxford University Press, Calcutta (Kolkata).

Hancock JF 2008. *Temperate Fruit Crop Breeding: Germplasm to Genomics*, 1st South Asian Ed 2018. Springer.

Hutchinson J 1959. *Families of Flowering Plants*, Vol I, II, 2nd ed, Oxford, London.

Keskar BG, Karale AR, Dhawale BC and Choudhari KC 1990. "G-137"- a promising clonal selection (pomegranate), *J. Maharashtra Agric. Univ.,* 15 (1): 105-106.

Khader A, Md JBM, Kulusekaran M and Muthuswami S 1982. Co-1-A Soft seeded selection. *South Indian Hort*., 30 (4): 298.

Kumar M 2015. *Fruit Breeding*: *A Modern Approach*. Bio Green Books.

Kumar N 2018. *Breeding of Horticultural Crops: Principles and Practices*, 3rd ed. New India Publishing Agency, New Delhi.

Levin GM 1990. Induced mutagenesis in pomegranate. *Refrativnul Zhurnal,* 10: 3399.

Mishra K, Mishra N and Chand S 2017. Pomegranate. *In: Fruit Breeding and Approaches*, pp.150-155.Published by International Book Distributing Co., Lucknow.

Misra KK, Mishra NK and Chand S 2017. *Fruit Breeding*: *Approaches and Applications*. Biotech Books, New Delhi.

Mukherjee SK 1953. Origin, distribution and phylogenetic affinities of the species of *Mangifera indica* L. *J. Linn. Soc. Bot.*, 55: 65–83.

Mukherjee SK 1997. Introduction: Botany and importance. In: Litz RE (Ed). The Mango: Botany, Production and Uses. CAB International, Wallingford, UK, Pp.1-19.

Naik BS 2022. *Horticultural Crop Breeding: Principles and Practices*. New India Publishing Agency, New Delhi.

Naik VB 1975. Seedling selection in pomegranate (*Punica granatum* L.) cv. Muskat. *M.Sc. (Agri) Thesis,* MPAU, Rahuri.

Nataraja K and Neelambika GK 1996. Somatic embryogenesis and plantlet from petal cultures of pomegranate (*Punica granatum* L.). *Indian J. Expt. Biol*., 34 (7): 719-721.

Nath N and Randhawa GS 1959a. Studies on floral biology of pomegranate1. Flowering habit, flowering season, bud development and sex ratio in flowers. *Indian J. Hort.*, 16: 61-68.

Nath N and Randhawa GS 1959b. Studies on cytology of pomegranate (*Punica granatum* L.). *Indian J. Hort*., 16: 210-215.

Nijjar GS 1977. *Fruit Breeding in India*. Oxford & IBH Publishing Co, New Delhi.

Pandey BP 1976. *Taxonomy of Angiosperms*, 3rd ed. K Nath & Co, Meerut.

Ponnuswami V, Kumar SM and Sundharaiya K 2015. *Blossom Biology of Horticultural Crops*. Jaya Publishing House, Delhi.

Ponnuswami V, Padmadevi K and Kumar SM 2012. *Botany of Horticultural Crops*. Narendra Publishing House, Delhi.

Ponnuswami V and Sumathi T 2017. *Breeding of Horticultural Crops*. Aavishkar Publishers, Distributors, Jaipur.

Purseglove JW 1968. *Tropical Crops*: *Dicotyledons* (Vol 1 & 2 combined),1st ELBS ed 1974, Reprint 1984. ELBS and Longman, UK.

Ramachandra RK, Maheshwar DI and Shoba N 2016. *Breeding of Spice and Plantation Crops*. Jaya Publishing House, Delhi.

Ramachandra RK, Nachegowda V and Honnabyraiah MK 2015. *Breeding of Fruit Crops*. Jaya Publishing House, Delhi.

Ray PK 2002. *Breeding Tropical and Sub-tropical Fruits*. Springer.

Rout G R, Mohapatra A and Jain SM 2006. Tissue culture of ornamental pot plant: a critical review on present scenario and future prospects. *Biotechnol. Adv*., 24: 531-560.

Roy D 2016. *Crop Evolution and Genetic Resources: Agricultural and Horticultural Crops*. Narosa Publishing House, New Delhi.

Roy D 2019. *Breeding of Fruit Crops*. Narosa Publishing House, New Delhi.

Samson JA 1986.*Tropical Fruits*. 2nd ed. Longman Scientific and Technical. 1986. pp. 216-234.

Shao J, Chen C and Deng X 2003. In vitro induction of tetraploid in pomegranate (*Punica granatum*). *Plant Cell Tiss. Organ Cult*., 75:241–246.

Sharma DK and Majumder PK 1988. Further studies on inheritance in mango. *Acta Horticulturae,* 231: 106–11.

Sharma G, Sharma OC and Thakur BS 2009. *Systematics of Fruit Crops*. New India Publishing Agency, New Delhi.

Shukla AK, Shukla AK and Vashishta BB 2004. Pomegranate. In: *Fruit Breeding: Approaches and Achievements*. pp. 245-246. Biotech Books, Darya Ganj, New Delhi.

Shukla Anil K, Shukla Arun K, Noor Mohamed MB, Singh A and Tiwari D 2020. *Fruit Breeding*: *Approaches and Achievements,* 2nd ed. New India Publishing Agency, New Delhi.

Shukla P and Misra SP 1982. *An Introduction to Taxonomy of Angiosperms*, 3rd ed. Vikas Publishing House Pvt Ltd, New Delhi.

Simmonds NW 1976. *Evolution of Crop Plants*, Reprint 1979. Longman, London.

Singh BD 2018. *Plant Breeding: Principles and Methods*, 11th ed, Reprint 2019. Kalyani Publishers, New Delhi-Ludhiana.

Singh LB and. Sturrock D 1969. Mango. *In:* Ferwerda FP and. Wit F (Ed), *Outlines of Perennial Crop Breeding in the Tropics,* pp: 309–27.

Singh SK, Patel VB, Goswami AK, Prakash J and Kumar C 2019. *Breeding of Perennial Horticultural Crops*. Biotech Books, New Delhi.

Srivastava KK 2009. *Systematic Description of Fruit Crops*. International Book Distributing Co, Lucknow.

Terakami S, Matsuta N, Yamamoto T, Sugaya S, Gemma H and Soejima J 2007. Agrobacterium - mediated transformation of the dwarf pomegranate (*Punica granatum* L. var. nana). *Plant Cell Rep.*, 26:1243–1251.

E-resources

http://mangifera.res.in/varieties.php

www.itfnet.org.